A PICTORIAL HISTORY
OF CANALS

A
PICTORIAL
HISTORY
OF
CANALS

D. D. GLADWIN

ILLUSTRATED BY J. K. EBBLEWHITE

B.T. BATSFORD LTD

LONDON

By the same author
ENGLISH CANALS
THE CANALS OF BRITAIN
THE CANALS OF THE WELSH
VALLEYS
VICTORIAN & EDWARDIAN CANALS
from old photographs
THE WATERWAYS OF BRITAIN
A Social Panorama

First published 1977
© 1977 D.D. Gladwin

ISBN 0 7134 0554 6

Filmset by Servis Filmsetting Ltd, Manchester

Printed in Great Britain by
The Anchor Press Ltd., Tiptree, Essex
for the Publishers
B.T. Batsford Ltd,
4 Fitzhardinge Street, London W1H 0AH

CONTENTS

BASICS

THIS IS not, and does not purport to be, a complete history of English, Scottish and Welsh canals. Aside from the physical limitations inherent in any one book, there were at their heyday more stillwater navigations than the number of photographs within these covers.

Again, I have not used a plethora of historical scenes depicting the inevitable horse-drawn narrow boats with their crews dutifully smartened up for the photographer. To a great extent these always seem unreal and to some degree are unfair to the reader. To quote an example – it is of little use utilizing a photograph of a 'Shroppie Fly' laden with china in 1910, when instead I can show an ex-Grand Union motor boat laden with coal – the latter can still be seen (however rarely) plying the same trade.

Furthermore even the use of the word canals in the title is to present something of a misnomer. The accepted definition of a canal is 'a navigable man-made water channel', and by this definition we would have to exclude navigable rivers which may or may not have had their channels modified by man. Instinctively the mind conjures up a vision of the Upper Thames, tamed and improved over many centuries. But how different are the Severn, once navigable to Welshpool, now (with a struggle) to Arley, but more commonly Stourport; the Forth, once connected by man-made waterway to the Clyde; or the Calder, which can rise three feet overnight. Bending the rules somewhat to compress the subject to handlable limits I have re-defined 'canals' as 'inland navigable water channels'.

The total navigable mileage extant at any one period has been calculated and re-calculated by many people, no two agreeing, even the Board of Trade claimed a figure of 2773 miles (4462 kilometres) in 1883 but increased this to 3320 miles (5342 kilometres) by 1898. An independent engineer, two years later, quoted the following statistics:

Owned by public trust	*927½ miles*
Independently owned	*1445¼ miles*
Guaranteed or owned by railways	*1333 miles*
Derelict	*118½ miles*
Of unknown ownership	*36¾ miles*
Thus giving a total mileage of	*3861 (6212 kilometres)*

The governing body on nationalized waterways, the British Waterways Board, claim they control 2000 miles, while a further 600 or so miles, in varying states of navigability, are under private, or at least non-BWB, control. The Basingstoke, once the property of one of Horatio Bottomley's companies, is now jointly owned by the Hampshire & Surrey County Councils, while the Neath falls under British Petroleum's umbrella. The Stratford (Southern) is a National Trust property. All three, and others, are run independently, however state-owned the holding companies may be.

The most obvious adjuncts to the basic water channel are locks which control or stabilize the levels, and reservoirs used to provide both an overflow to absorb spare water and to feed the channel in times of shortage. Obvious too, especially for the fretful motorist, are the bridges which cross the channel, whether brick, concrete or stone built; round or square of arch; swivel, swing, bascule or unbalanced draw.

Less obvious are aqueducts, culverts, weirs, wharves (often, sadly, decayed), and innumerable land-drains taking off superfluous water from land, road or housing estate. Visible, albeit often neglected, are waterside factories, dwelling-places and warehouses, but far from apparent to the casual onlooker are the minutae that, when all is said and done, make a navigable water channel just that: the keystone of a bridge arch, the lock gate collar, brick, stone or clay banks, a boundary marker or milepost, even the iron guard on the approach to a bridge.

Add to this infinite variety the fact that too often the ostensible engineer of a waterway, whether Brindley, Rennie or Telford, was directly involved only in surveying the line to be taken, the physical building being supervized by a Resident Engineer – covering 30, 40 or 50 miles on horseback – having under him contractors of varying ability, each with his own ideas, and it becomes obvious the permutations of style, finish and appearance are innumerable. Throw in for good measure the varying character, and even colour, of the materials available and a goodly number of ironmasters tendering for any metal parts, and it will be seen that any book, however large, can only skim the surface of the subject.

The principal problems that arise in sorting out available material are twofold. Photography became a popular occupation only in the latter decade of the 19th century and even then photographers were rarely interested in industrial scenes. Were these not immutable? The sun could never set on England's glory and therefore her industry would never change. For my own part, I, too, fell into this trap in the 1950s when although I lived alongside a waterway and worked boats (with intermissions elsewhere when bankruptcy loomed) I never thought that Hednesford Basins or Broad Street Wharves could ever be other than full of boats, but instead, clutching a box-camera, plodded around the Wiltshire and Berkshire, Grand Western and similar disused waterways on the basis that these would soon be gone for ever. This proved doubly fallacious, both Hednesford and Broad Street are, or will soon have, gone for ever – following the 7

Triptych of engineers. Leader–Williams, William Jessop, Thomas Telford.

boats – while the Grand Western and the Wilts & Berks are being revitalized by the growth of amenity traffics.

Worse still today, we can no longer photograph scenes that were here yesterday. A brief decade ago quiet country areas looked more or less as they must have done a century before but now – yesterday, today, tomorrow, either unconsciously by the building of a new housing estate, or consciously by wanton destruction of an old cottage on behalf of some nationalized or mult-national group, the old way is destroyed for ever. How far it is gone can be seen by perusing the photographs in *Victorian & Edwardian Canals from old photographs* (B.T. Batsford Ltd. 1976) and looking at those scenes today.

Three examples – among hundreds – spring to mind: Telford's great iron Galton Bridge spanning the Birmingham Canal Navigations Main Line gave pleasure to thousands of eyes as cruisers passed it by, now the new tunnel built to accommodate the road traffic obliterates the scene. For all that the Anglesea Branch Canal carried a heavy tonnage of coal until 1963, it was described in a guide published in 1968 as 'predominantly rural'. Urban is the word to use in 1976, as housing estates appear on its banks and, unhindered by such trivia as planning consents, mounds of scrap cars, lorries, tyres and rusting metal proliferate. The triumvirate is completed by recourse to the old, but true, adage 'beauty lies in the eye of the beholder'. Is the new, tidy, wharf at Tardebigge (Worcester & Birmingham Canal) better than the old? Clinically tidy it may be, but where is its life?

Change, if not wanton destruction, occurs on waterways owned or operated by the British Waterways Board under the guise of maintenance works. Not that change is such a bad thing in any field, but this change must be sympathetically planned, carried out by men in harmony with the subject, and the materials used must blend with, and preferably enhance, whatever the subject may be.

With the fantastic erosion caused by poorly designed, over-engined and irresponsibly handled cruisers growing worse daily, piling of the banks has to be carried out by the Board's workmen. This is in itself a sensible thing, and the necessary backfilling of the gaps on the towpath with mud dredged from the canal fulfils two purposes – making channels, or roads, for both pedestrians and boats. Upon this mud wild-life can grow, Shakespeare may well have looked upon the Avon when he wrote:

> *I know a bank whereon the wild thyme blows,*
> *Where oxlip and the nodding violet grows;*
> *Quite over-canopy'd with luscious woodbine,*
> *With sweet musk-roses, and with eglantine.*

In addition it is easy to find buttercups, daisies, speedwell, vetch and tansy but only on a disused waterway. These 'weeds' undeniably add charm, even beauty, so why are towpaths chemically sprayed to destroy the growth? It is claimed, justifiably perhaps, that mowing is expensive in terms of man-hours, but what are the long term effects of spraying? If the Board's management are so whole-heartedly in favour of spraying why use an inhibitor only – why not sodium chlorate or any other 'total' killer and do the job properly? Why kill the broad-leafed flowers as they do and leave rank 2-feet high grass? This is, admittedly, an emotional argument and given sufficient weight of public opinion the practice of spraying could be stopped tomorrow, but regard – when a lock wall is rebuilt now, as it may have been previously (twice, or even three times) why are the old attractive, weathered sandstones discarded, broken up to be replaced by concrete? This is irreversible. Irreplaceable, too, are the old cast or wrought iron bridges, products of Dallam, Toll End or Horseley Ironworks, now being steadily demolished to be replaced by relatively featureless concrete constructions.

Conversely though, credit must be given to the men or women who endeavour to build museums, often from the proverbial grain of mustard seed, no longer 'fossilized' relics in a gaunt Victorian building, but instead living entities. Therefore, in this survey, I have endeavoured to treat the waterway network as an oversized Industrial Archaeological project; and tried to show what can be seen at this stage of the excavation. Like all good sites, though, in one corner a modern concern is functioning on the old foundations – the Aire & Calder Navigation together with its contiguous waterways still transports between two and three million tons per annum and provides a welcome contrast.

It is obvious that in a book of this nature I have had to draw upon the resources of many friends and colleagues and it is difficult to

express my gratitude to them, both for the loan of many illustrations and for readily granting a waiver of their copyrights. Photographs are acknowledged individually, but for the copying and printing of many I am obliged for the meticulous care taken by Frederick W. Jones of Solihull. Proof reading, as ever, has been in the capable hands of Hugh Barker, while J.K. Ebblewhite has enlivened both this and many other books with his engravings. My wife was both typist and overlooker. Finally in drawing on a collection of 20,000 photographs it is just possible that acknowledgements may be omitted or wrongly given; I would pray forgiveness for any errors and if advised do promise to make amends upon reprinting.

ONE OF THE greatest charms of any waterway is the long pounds, or lengths, between locks. The appearance of these is dependent on a number of geological and man-made features; and these in turn are influenced by the techniques of the engineer concerned.

The earliest waterways of which we have cognizance are the Caer and Foss Dykes, dug by the Romans to ease transport of grain from the hinterland to their forts and towns at Lindvm (Lincoln), Cavsennae (Ancaster) and Dvrobrivae (Water Newton). Typical of their techniques, they were straight as were, in the main, those of that 17th-century genius Vermuyden and, within the limits imposed by landowners two centuries later, those of Telford, including his last and greatest, the Birmingham & Liverpool Junction Canal, now the main line of the Shropshire Union.

In the interregnum came the bulk of the waterways still open today. These were built to a so-called 'contour' pattern, keeping as near as possible to a fixed level, irrespective of any meanderings that might result. These twists and turns were indeed often turned to use, as small villages could be served, although through transport was delayed. This, within natural limits, was quite a satisfactory state of affairs for about 30 years, until two external factors had to be taken into account. The first was that, due to the ease of movement available on a waterway, town populations and manufactories grew apace, both requiring raw materials to feed them and an enlarged market to absorb finished products. The second influence was that of turnpike roads. Even in the 1820s we find commentators complaining that goods could be transported by road quicker than on a parallel cut. It is a salutory thought that in England and Wales alone there were already by 1829 no less than 20,875 miles (33,588 km) of Turnpike – while at best navigable waterways totalled only 4000 or so.

At this juncture there arose a dichotomous situation. The proprietors of the older (and in the main most successful) waterways, were suffering from a distinct hardening of the mental arteries. Not only had the prime cost of their waterways long since been paid off but they had grown to like the rattle of the cash-till as their $37\frac{1}{2}$ per cent (Trent & Mersey, 1821), 58 per cent (Erewash, 1824), 32 per cent (Oxford, 1833), 44 per cent and 61 per cent (Coventry, 1824) dividends were paid out.

As long as they stayed away from the towpath they were unable to hear the swearing of the boatmen as they struggled along the Main Line of the Birmingham Canal, described by Telford as '... little better than a crooked ditch with scarcely the appearance of a haling-path, the horses frequently sliding and staggering in the water, the haling-lines sweeping the gravel into the canal and the entanglement at the meeting of the boats incessant; while at the locks at each end of the short summit

CHAPTER ONE
THE BROAD CANVAS

crowds of boatmen were always quarrelling, or offering premi-
ums for a preference of passage, and the mine owners, injured by
the delay, were loud in their just complaints'. Nor, as they did
not open the day's post, were the proprietors in any danger of
reading the vociferous complaints of shippers and receivers of
freight that was either mislaid, delayed or damaged. Eventually
prodded by the Company Secretary and/or influential shareholders, like the Wedgwood family, an engineer would be called
in to take the necessary steps to straighten some twists and turns.
The prime example of this treatment is the case of the Oxford
Canal, where one half – the Northern between Braunston and
Hawkesbury – was taken in hand between 1829 and 1834 while
the Southern was left as built. In other cases the 'track' might
only suffer minor changes, but the locks would be widened – or
even duplicated – to allow for larger craft, or at the very least,
less congestion. On the other hand available traffic was so excessive that stagnation threatened the livelihood of all, and a few
new 'direct' canals were built at vast cost and (in the upshot)
too late.

The contour versus direct canal argument could backfire. The
Rochdale and the Calder & Hebble canals jointly provide a broad
canal of some 54 miles from Wakefield to Manchester and incorporate 124 locks. The Calder & Hebble, the Huddersfield
Narrow and the Huddersfield Broad canals plus a part of the
Ashton provide a similar service albeit of 42 miles. But this latter
route, which had a total of 119 locks and into the bargain a three
mile tunnel, was a narrow canal taking only ten (summer) to
twenty (winter) ton boats. The Rochdale for all its extra length
flourished, while the Huddersfield Narrow failed.

Naturally there were exceptions to a very general view. The
Caledonian canal in Scotland was almost a true contour canal
but was as direct as could be engineered, bearing in mind that the
supply of water for any waterway is the critical factor affecting
the flow of traffic; and it has been shown that if, for example, the
Southern Oxford had been straightened to the extent of its
Northern half, boat traffics would have had to cease in the
summer months for the twisting loops around Napton and
Wormleighton at least act as reservoirs. The bulk of the Welsh
Valley canals could follow only one line for without altogether
excessive costs in blasting and labour they had no option but to
wind and twist along a shelf.

But there were places where canals were physically and
economically unrealistic and various attempts were made to
close the gaps, mainly by way of tramroads. The precise mileage
of these will probably never be known, as some were ephemeral,
serving a colliery for a year or two then being uprooted to a new
location. Some indication of their extent can be gauged from the
fact there is known to have been in use in South Wales, between

1780 and 1850, no less than 326 miles (525 km), all feeding to and receiving traffics from a mere nine navigations.

Other tramroads bridged the gap in the Grand Union canal while Blisworth tunnel was being built, connected the Cromford and Peak Forest canals and the Leeds & Liverpool with the Lancaster canal. Within the Black Country district tramroads abounded even until relatively recent times, although, as in many places, the most vital were made suitable for steam haulage.

As the years rolled on – and it must be remembered most waterways are over 175 years old – changes have occurred in their appearance, partly due to man's influence and greed, and partly as nature does her best to hide the scars – something that she will never do for motorways.

As miners worked their mole-like way underneath the land, so in the fulness of time 'tocky-bonks' or waste-heaps grew, and equally as inevitably the land subsided. Rather than have the canal resembling a switchback, maintenance men would raise the banks, and now often a channel ten or twenty feet above the ground appears to follow an utterly purposeless serpentine course. It is not entirely coincidental that waterways so treated are among the deepest in the country! The converse can apply where a canal passes through a deep wood cutting as years of leaf-mould, and soil erosion, can gradually fill it in. In the case of the Stratford, built as a small-boat waterway only three feet deep, nearly a century of neglect has made subsequent cleansing difficult, the original depth rarely being available. Curiously, during one dredging operation at about two feet depth blue clay was lifted by the grab, followed by bone-dry peat. Here, presumably, the weight of nearby flats and a road was making the bottom buckle.

Erosion from the surrounding land is another cause of change, and nowhere more apparent than in the Fens; it is difficult to visualize the Holland, Forty-foot or Hobhole Drains as the land drainage channels which once they were.

Locks, in the main, seldom raise or lower a canal more than six to nine feet, and even where a flight of locks – the 29 of Tardebigge (Worcester & Birmingham canal), the 21 of Hatton (Grand Union canal) or, again, 21 at Wolverhampton – occur, usually a twist in the canal, or a building, obscures the panorama. To some degree it is more satisfying to the soul to stand on the towpath of the Huddersfield Narrow at Milnsbridge between 'dark, satanic' mills (cleaner nowadays) and see the waterway as a green finger bringing the countryside to a city.

3 *Opposite* Towering loads of hay on 'stackies'. A traffic far older even than canals, it was carried on a very primitive form of sprits'l barge. These probably worked from the Essex creeks and for this reason the captains had their own nickname — the 'Reed Buntings'.

4 Perhaps the archetypal illustration of a dream canal. At the close of another serene, leisurely and (inevitably) sunlit day one man, pausing only to pass words with the lock-keeper, leads his horse away to the stables, while his mate steers the laden boat neatly into the lock, prior to tying up and making their way to the hostelry. To the right steam pours from Batchworth Mill and a laden wagon goes on its way. Grand Junction Canal 1819.

5 *Left* In this painting we have all the rustic glory of a country canal. A laden boat with the good lady talking to the squire while two hinds gaze at the scene. River Avon, *c.* 1890. *E. Fenton*

6 Another historic traffic is that of 'round timber', in this case depicted on the River Avon at Stratford, *c.* 1795. Evening calm, the wind has dropped, so the steerer quants his way towards the wharf.

7 Although built between 1804 and 1806 to give 'a defensible line of great strength' against the Napoleonic armies at a cost of £200,000, the Royal Military Canal had already sunk into slumber by 1820, the 'gondola' being hardly a warlike provision.

8 In its heyday Brimscombe Port was a major inland port, forming the junction between the Stroudwater Canal (and River Severn) and the Thames & Severn Canal (which gave access to London). Engraved about 1823 this scene appears to show works continuing, although by then the Port had been in use for 34 years.

9 Stoke Bruerne on the Grand Junction Canal around the turn of the century. The whole scene is radically different from that of today, being then a commercial canal in an era that feared not road traffic. The arm to the left is now infilled and used as a car park, the tall building, a mill, is now the Waterways Museum.

As the years ground on, canals met and were either absorbed into or fought with the railways. Weakened by this, narrow canal carriers were in a poor position to compete with road transport; all the more so as the majority of Governments would (until recently) have gladly filled in the channels.

10 The Caledonian Canal, although busy enough even today, by its sheer size defies any attempt to make it look industrial. The lochs, indeed, make the whole waterway rather improbable and vaguely reminiscent of the Rhine. Castle Stacker, Loch Linnhe in 1957, now restored to habitation. *G. Davies*

11 Wolverhampton 21 locks. An empty train rattles by a lonely pleasure boat.

12 A lorry sneers down at the closed length of the Dudley Canal, Mucklow Hill.

13 Canal Head, Aberdare in 1971. This was the point at which iron was trans-shipped from tram to boat after being conveyed some three miles (5km) from Hirwain. Closed to navigation in 1900, the bulk of the waterway was infilled and converted to a road in 1923. *G. Davies*

14 Wyrley & Essington Canal dreaming quietly near Huddlesford Junction.

15 It is difficult to realize this is a derelict canal; the piles in the foreground are all that is left of a busy wharf. Craigton Bridge, near Whitequarries, Edinburgh & Glasgow Union Canal. The whole line was abandoned by Act of Parliament in 1965. *D.G. Russell*

When canals die – or more accurately are killed –
their end can be lingering. This is fortunate in some
ways as, today, many moribund waterways could
still be revived.

16 *Previous page* Market Weighton in 1969 — more a rubbish tip than a water channel.

17 and **18** This is the dreaded Bishboil. Once a part of the Dudley Canal it was superseded in the 1850s when the line was straightened out, although still remaining in use as a deviation and for local traffic. Now truncated it is slowly being infilled. The bridge is virtually a standard pattern for that part of the Birmingham Canal Navigations.

19 Netherton Branch Canal with Cobb's Engine House to rear. An open ended 'Watt' beam engine was in use here until 1928 by the South Staffs Mines Drainage Board, pumping from the mines to the canal. The engine was subsequently sold to the Henry Ford Museum in the United States of America.

Some navigations have been tidied up, but while slag heaps and marshes are unattractive it seems difficult to find the balance between that and sterilization.

20 Anglesea Basin (Birmingham Canal Navigations) is now cleared of these 'Joey' (day) boats which were abandoned there when the coal runs to the Central Electricity Generating Board power stations dwindled away in the 1950s and 60s, but is absolutely featureless and a stamping ground for hooligans.

21 Worcester & Birmingham Canal at Kings Norton; its junction with the Stratford Canal (right) showing the quondam canal manager's house.

22 At the junction of the Wyrley & Essington and Lord Hay's Branch the iron bridge shows (foreground) the wear of many ropes — each passing over, by its friction, made the grooves a little deeper — but how many must have passed?

23 When, on October 30th, 1934, the Duke of Kent, seen here on the experimental boat 'Progress', opened the rebuilt Hatton Locks (Grand Union Canal), it was anticipated that the substitution of wide locks and wide boats for their narrow forerunners would bring about a new era in water-borne transport from London to Birmingham. The former had little chance to prove themselves prior to the outbreak of war, the latter were failures. Interestingly, 'Progress' had a German (Junkers) diesel engine . . . a sound heard later in different circumstances. *The Geographical Magazine, London.*

24 A privately owned push-tow tug hurries purposefully along the Wyrley & Essington Canal near Aldridge. Once a 'contour' canal — following a level, however serpentine, course — mining has caused the land on the right to subside, leaving the waterway high above.

25 The Calder & Hebble at Brighouse could, with a little imagination, be mistaken for a commercial waterway. At least, even on a gloomy day, life is apparent.

The treatment of the track itself is as variable as that of the surrounding land. The canal may be resuscitated or may be neglected.

Over the years, perhaps encouraged by relatively clean water, house- and business-owners have evinced a pride in their water-side frontages. Others, unfortunately not caring, present a very different façade.

26 This scene although photographed as recently as 1967 is pure history. The canal, in order to give more mooring space, has been widened immediately behind the clubhouse. Govilon, Brecon & Abergavenny Canal.   *G. Davies*

27 Gargrave, Leeds & Liverpool Canal.

28 Swan Village, Ridgacre Canal.

29 The Worcester & Birmingham Canal at Blackpole Lock (No. 9) one misty November's day is gloriously atmospheric presenting, once again, an idyllic scene. Rarely, alas, do pleasure cruisers venture out at such a time!

30 A direct contrast with industry is this animated summer scene. Posed though it may be, it is far removed from dereliction and decay. *British Waterways Board*

CHAPTER TWO
ARCHITECTURAL GEMS

THE BASIC structures to be found alongside any stillwater navigation can be divided into a number of groups. Service buildings include lock- and bridge-houses for the accommodation of workmen, plus, for a similar purpose, odd terraces usually having a collective name 'Tug-row', 'Tunnel Houses' or 'Wharf Cottages'. These dwellings have varied in their condition from the idyllic, trim, neat, honeysuckle-over-the-door pattern to those described succinctly enough as being 'in time of flood, the door half open and the floor all awash; tenant had fled'. Other, larger, premises were built for the benefit of managers and 'others of rank'.

Each canal had its own conglomeration of workshops; a lock-gate would arrive in the yard as a tree trunk and each operation would be carried out within those precincts. The sawyers would rough-cut the timber, the carpenters saw, plane, adze and trim the balance beams, rails, bumpers and face-planking together with any other requisite woodwork, while the blacksmith and his mate would produce the necessary ironwork. The whole assembled, the gate would be loaded by means of a hand crane onto a maintenance boat, which might itself have been built in the self-same yard. Bricks, lime, horse-dung (for making chalico, as docking was another task), tar, nuts and bolts (blacksmith made), lead, boats-lines, cloths, hackers, sickles, scythes, brushing hooks and 'ever-so-many' other items were to be found in the yard or stores. Vital, too, for the easy running of any waterway was the dry- or side-dock, and stabling for horses. Some yards – Sneyd on the junction of the Wyrley & Essington and the Sneyd & Wyrley Bank canals makes an example – were linear in form, others (Hartshill on the Coventry and Icknield Port on the Birmingham Canal Navigations) make a pleasant block; although putting cottages, Inspector's house and workshops together tends to lead to a parochial form of life.

Including farmers' stages for the loading of milk churns and occasional discharge of manure it has been estimated that on average there was one wharf per quarter-mile of waterway. Virtually every village had one – some a simple wooden structure, long since decayed; others a more substantial affair of brick, with an iron bumper to guard against wear and tear. These may yet remain, perhaps serving as moorings for pleasure craft or, more likely, for access is only suitable for horses and feet, lurking underneath grass, forgotten except by a few locals. Some of these wharves incorporate a range of buildings – a few houses, stabling and, with luck, even a pub! Often a cottage will have a bow window in front, and this could well denote that once this was a toll house where the charges were collected from the boatman. Even more rare were gauging stations – in a few cases these would have a counterbalancing mechanism allowing the boat's weight to be measured and its laden draught calculated. One of these gauging stations, Midford on the Somerset Coal

Canal, was almost grandiose, complete with portico. Virtually every wharf had its crane, wood or iron, manumatic in operation, but these are among the artifacts that are fast disappearing or being modified, that at Hanbury (Worcester & Birmingham) having its pedestal incongruously painted to imitate a lighthouse. Variations on a wharf-theme can be rung almost ad infinitum – old tug winding holes (i.e. where the boats were turned) will have, somewhere nearby, the relics of a hovel – for the men – and a coal stage – for the tugs. Others may still have the stumps of cattle or sheep pens still standing, or of chutes for loading, or be on an arm of the waterway with, above, rail tracks still discernible through the under and overgrowth.

The next stage in the growth of a wharf is the presence of warehousing facilities. The scale of these, like everything else on canals, varies according to demand although so parsimonious were the Company of Proprietors of the Birmingham Canal Navigations that, rather than build extra storage space (and this at a time when their dividends were in excess of 50 per cent) they utilized extra craft as floating warehouses, to the detriment of goods, both from the damp and light-fingered souls. The vital necessity for proper covered space can be gauged from the complaints of both consignees and consignors, and it is as well to recollect that at least $8\frac{1}{2}$ million tons traversed the Birmingham Canal Navigations in 1898, while as late as 1905 722,000 tons were being handled on the Staffordshire & Worcestershire canal. Assuming that only a half of this tonnage needed storage at one end or the other and that, again, only a half of this was perishable, this still leaves a fair amount of warehousing space to be found, not only in weight but also by the cube. A narrow boat carrying 25 tons (25.4 tonnes) has a usable space roughly equivalent to 1500 cu. ft. (43 m^3), while barges are proportionately larger. As an adjunct to the requisite storage space, a canopy was desirable to avoid wetting goods as they were discharged; these are still extant, often in backwaters of canals where no boat dare venture for fear of its propeller ('the blades' in canal parlance) being fouled by the detritus of civilization. Others are distinctly 'spooky' as they loom out of an autumn mist or winter's fog, but the bulk are forlorn waiting for the developer's hammer to strike. Occasionally, although no longer served by canal, they are adapted to the purposes of road-haulage; but this always seems to be a temporary expedient, erasure and rebuilding (if not arson or vandalism) being programmed. Latter day waterside erections include silos, mechanical coal loading staithes and even overhead loading gear.

Many of these warehouses will still show (albeit bricked up) the holes or archways used for loading or discharging, while above may be an iron pulleywheel or hoist; this may incorporate a sliding arm, most likely to be an iron or, if relatively modern,

*Disused warehouse with canopy. Wordsley, Stourbridge Canal.
Milepost. Church Lawton, Trent & Mersey Canal.*

steel joist or, more sophisticated yet, a swinging-arm crane, or series of cranes, more or less firmly bolted to, or through, the wall. Again, variety is prevalent, these cranes may be manually operated from within the building or by chains below, or steam or electrically driven or – very intricate this – by hydraulic (water) pressure.

However, none of the foregoing would be of much use without the factory whence the goods originate. A few factories antedate the waterway but most typical are those built during the heyday of canals, say 1770–1835. Built not only for the advantages of easy transport but also because water, still necessary even today, was then vital. Machinery was steam fired, engines were water cooled and some processes, especially in the North, required vast quantities of the 'essence du canal' to be successful. Despite the condition of some waterways demand remained high for many years, indeed by the end of the nineteenth century for many shareholders the income from water sales provided their only chance of a dividend.

In the 1960s a scheme was brought forward wherein three possibilities for the future of non-commercial waterways were propounded. The first was closure, the second conversion to a water channel and the third retention for use by pleasure craft. The third was chosen, and it was found that in many cases closure was neither physically nor financially feasible; as the value of the channel for water supply (including use as a drinking water reservoir) far outweighed any possibility of closure.

The least obvious of the authentic canalside structures, or at least artifacts, are the mileposts, mooring stumps and boundary markers still to be found here and there. These grow fewer daily as many are easily transportable and serve to prove – for a while at least – that some deranged vandal has been to a certain canal; until he/she grows bored with and dumps the item. Other

34 accessories are finding a ready market both at home and abroad,

following a trend set by collectors of railway relics. Mileposts are all too conspicuous – the ornate disappearing far faster than the plain! Boundary posts are, seemingly, only of value if marked with the name of the company (G.J.C.Co., S. & W.C.Co., etc.), mundane GWR markers remain inviolate. Mooring stumps are being destroyed for two reasons, the first is a present day desire for standardization, hence replacement by concrete toadstools, and the other, I fear, is simply that they are there.

A final structure visible and known to many is the canalside hostelry. Once known for rich beams, blackened ceilings, dark oaths and darker beers, many have now changed as the result of a swing to road traffic and provide white ceilings, pale faces, clean plastic surfaces and ersatz beer. Still, if this is to modern taste, so be it – and at least clean suits can be worn by Mr and Mrs Cruiser!

'Tunnel House', Sapperton, Thames & Severn Canal.

Canal buildings are infinitely variable, in purpose, design and location and therein lies their charm. Bridge, lock and toll houses are not the least important.

34 The way down to the Sheffield & South Yorkshire Canal at Rotherham. Not so heavily used now, in bygone days generations of bargemen clumped their way to and fro.

35 Gloucester & Berkeley Canal, Hardwicke Bridge. Originally designed under the guidance of Thomas Telford, these cottages — there is one for every bridge — are elegant, if inconvenient, homes for the bridge-keepers.

36 Aire & Calder Navigation. Stanley Ferry lockhouse, now a depot for T. Fletcher & Sons. The wharf received 50,000 tons of grain and 20,000 of paper and timber each year. *P.L. Smith*

37 Another lockhouse, this time situated at the oddly, but melodically, named Bumble Hole Lock on the Staffs & Worcs Canal. The television aerials are the only major additions or alterations on this Georgian building.

38 Calder & Hebble Canal. This lockhouse at Brighouse represents a pleasant oasis in a predominantly industrial neighbourhood; the edge of a mammoth gasholder is visible to the right.

39 There are however hazards in inhabiting any canalside dwelling. Detail of wall flood-marker from left-hand side of figure 38.

40 The tollkeepers 'hovel' at Groveland Bridge, Tividale, on the Netherton Tunnel Branch Canal, had altered little over the century since its building in 1848. The 'stop bars' whereby the boats were prevented from moving until they had paid their dues, are clearly visible to the left of the photograph.   *Dudley Public Libraries*

41 Junction House at Marple where the Macclesfield and Peak Forest canals make their union. Once the home of the engineer, this was still trim and little changed in 1966.

Calder and Hebble Navigation Office,
HALIFAX, March 8th, 1816.

ROBBERY.
FIFTY Pounds Reward.

Whereas

The WAREHOUSE belonging the Company of Proprietors of the Calder and Hebble Navigation, situate at Cooper-Bridge Wharf, near Mirfield, was broken open and feloniously entered by some Villain or Villains in the Night betwixt the 5th and 6th Days of March Instant, and the following Pieces of Cloth taken out of a Bale, deposited in the said Warehouse, marked B in a Diamond, Viz.

1 Piece Drab, marked N°.			1797,	measuring	36	Yards.
1 Do. Blue	Do.		1817,	Do.	$36\frac{1}{4}$	Do.
1 Do. Do.	Do.		1820,	Do.	$36\frac{3}{4}$	Do.
1 Do. Do.	Do.		1818,	Do.	$37\frac{3}{4}$	Do.
1 Do. Olive	Do.		1742,	Do.	39	Do.
1 Do. Do.	Do.		1739,	Do.	$40\frac{1}{2}$	Do.
1 Do. Do.	Do.		1743,	Do.	$40\frac{3}{4}$	Do.
1 Do. Do.	Do.		1744,	Do.	41	Do.

A Reward of FIFTY POUNDS

Is hereby offered to any Person or Persons who will give such Information as shall lead to the Conviction of the Offenders; and if any of the Perpetrators or Accomplices in the said Robbery, will inform of the other or others of them, he shall in like manner be entitled to the said Reward, and every Means used to obtain his Majesty's Pardon.

BY ORDER.

William Norris,

Jacobs, Printer, near the New-Market, Halifax

42 and **43** Leeds & Liverpool Canal warehouse, dwelling and Toll Office complex at Barrowford, still bearing in 1975, quite clearly, its pre-nationalization lettering 'Leeds and Liverpool Canal' over the doorway. Also shown is a detail of the stonework.

44 Theft from warehouses was not uncommon, as the offered reward indicates. The penalty was high — transportation at least — for what we now seem to regard as a normal way of life. *South Yorkshire Industrial Museum*

45 Wappenshall Junction, Shrewsbury Canal, 1971. This magnificent structure was clearly purpose built with the boats being able to load or discharge under cover, having made their entry through the archway. *A. Martin & D. Sumner*

46 This particular warehouse, now occupied by a boat hire-and-building company has been cleaned and refurbished, although thankfully the crane remains in situ. Sowerby Bridge, Calder & Hebble Navigation.

47 In total contrast is this view. Although at the time Great Bowden Hall was photographed (1969) it appeared somewhat unkempt, it is now rejuvenated. The ice on the Market Harborough Arm of the Grand Union Canal has also melted.

48 The Pocklington Canal never quite made it to the town, terminating at this warehouse at the foot of the hill. Lonely and forlorn as it may be at least it remains substantially intact and occupied.

In the heart of an industrial town buildings are prone to fall to the blow of a developer's hammer; the desire to make money still remains paramount in an ostensibly welfare-orientated society. Whether they are to remain in situ or to be replaced by some cubist's nightmare is a decision that lies sololy in our hands; supine apathy is the developer's friend.

49 *Previous page* Aire & Calder Navigation, Wakefield Depot, 1954. *P.L. Smith*

50 Photographed around the turn of the century, the quay at the Spalding Gas Works is being reconstructed to allow for the discharge of coal from Humber Keels. The ornamentation on the small keel, foreground, is traditional, as are the barrel winches. Obviously at this time the Welland was navigable thus far but four years later a guide states that 'the river is only navigable for paying loads on Spring tides'. *L.A. Edwards*

51 Gloucester Docks in all their glory around 1910. An impressive variety of craft, not excluding the ubiquitous narrow boat. To add to the interest, on the left stands a representative wagon of the Midland Railway and adjacent thereto a 'flat-bed' cart. *The Water Folk Ltd.*

52 Birmingham Canal Navigations. Tangye's Cornwall Works as publicized in 1884. Presumably the engraver went up in a balloon to get this view, the inevitable (almost compulsory) smoke did not, fortunately, obscure the lines of the buildings. The use of sailing craft (foreground, right) seems rather optimistic in view of the notoriously low bridges on the Birmingham Canal Navigations. *W.K.V. Gale*

Lest we forget, and in this hedonistic age there is a real danger we may, canals had – and still have – only one main purpose in life, the carriage of goods; the heavier the better. It was because of this ability that industry gravitated to their banks.

53 Although there are a number of illustrations of 'dark satanic mills', none seems more impressive than this scene on the Calder & Hebble at Brighouse, photographed in 1975.

54 Birmingham Canal Navigations, Priorfields, Coseley. Priestfield Furnaces were in full working order when photographed c. 1911. The boats are just visible (foreground) but are dwarfed by the buildings.   *W.K.V. Gale*

55 A gaggle or flock of 'Joey' boats lie underneath the shadow of Alfred Hickman's Spring Vale Blast Furnaces in the early 1920s. To the right the underlying cause of many day boats having crumpled sides is all too apparent. The rather surprising thing is that these craft had a life of 30–50 years, although like the Irishman's knife, very little was left of the original towards the end! Traffics carried included coal and iron ore in, and large quantities of pig iron produced in the furnaces (and seen in stacks on the left) went out, both locally and much further afield. The furnaces in the background (numbered left to right, 4, 3, 2, 1 and 5, 6) were reduced to three by Stewards & Lloyds shortly after their takeover in 1920. Again modernized in 1954, the whole new complex — virtually nothing remains of the items shown here — may be seen alongside the Birmingham Canal Navigations Main Line.   *W.K.V. Gale*

Whether in rural, urban or industrial surroundings, the canal-side hostelry was and (here and there) still is a vital part of the scene.

56 'Ivinghoe Ales, Stouts & Cyders' were all for sale at the Loco Hotel, Old Wolverton, around 1910; wharves, stables and, presumably, boats all lie behind the bridge. Push-bikes, rather than cars, represented road transport interests. *P. Garrett*

57 The Navigation Inn, Bridge 79, Maesbury Marsh on the Montgomery Canal in 1967. With, alas, the closure of the canal in 1944, the trade of the pub has had to be reorientated towards the road. *G. Davies*

58 The 'Water Gentian' of the British Waterways Board's hire fleet descends Buckby (or Whilton) Top Lock, Grand Union Canal. The dog, if not the person steering, would clearly enjoy a pint in the 'New Inn'. The rather odd painting at the cabin end is a very crude derivation of that once found on 'traditional' narrow boats. *British Waterways Board.*

59 Not a pub, but on the Caledonian Canal, The Ship Stores awaits you at Laggan Locks. Hospitable people, inhospitable countryside. The windlass (foreground) is used to pull vessels to the quay. *G. Davies*

60 Chapels once abounded along the banks. Here at Oldbury this one has outlasted the so-called 'infilled' canal.

61 Not quite what it seems. Co-operation between all interests, including anglers, has led to this becoming the main loading wharf for domestic coal traffic in the midlands.

62 Whether these bars will last much longer is very doubtful; the whole area is depressing and would, for once, be improved by development. Savile Town, Dewsbury Old Cut, 1975.

Not only buildings, but the smaller items make a waterway of such value to the industrial archaeologists. Alas, a small percentage of 'enthusiasts' cannot leave things alone, but in their enthusiasm must up and away with any even remotely portable items, and for this reason suffice it that: —

63 is somewhere West.

64 is somewhere North.

65 is somewhere in the Midlands.

66 is somewhere South.

67 Symbolic? Sheffield & South Yorkshire Canal, Ferrybridge Power Station.

CHAPTER THREE

LOCKS

'THE PENT WATERS, now freed, roar and rush joyous in their new-found freedom; then all is still and at peace.' A rather fanciful description of the egress of water from a lock is nevertheless accurate enough inasmuch as the water does rage, although an engineer would sooner mutter about water pressure being exerted through a too-small hole! Who, precisely, invented the pound lock is unknown to most mortals; the Egyptians for example built a canal during the 18th Dynasty (1335–1289 BC) between the Bitter Lakes and the Red Sea; in China one Emperor, Shih-Huang-Ti, is supposed to have cut one around 222 BC from the Yangtse River 600 miles towards the North; the Romans, those great hydraulic engineers, built innumerable stillwater navigations including the first in Britain, the Caer Dyke, and the Russians a fair mileage some 800 years ago. Any of these, with the sole exception of the Caer Dyke, could have had pound locks similar to those in use today. However, research has led to many possibilities emerging and the belief that such locks are relatively recent. The inventor may have been one of the brothers Dominico (c. 1481), Leonardo Da Vinci (c. 1519) or John of Ostrogotha (c. 1600). It is, however, certain that such locks were common-place in France during the middle 17th century. In Britain 'flash' locks – in effect only a movable dam – were used on many early river navigations, but in general were unsuccessful to the ruin of their promoters' wealth and, too often, health.

Within the land encompassed by the United Kingdom, the cutting of canals equipped with pound locks may well have commenced in 1563. This waterway, the Exeter Ship Canal, was completed three years later. It seems matters then rested until 1730, when work started on the Newry Canal. Taking twelve years hard work, this in turn led to the Sankey Brook being duplicated with a stillwater navigation (the St Helens Canal) rather than the stream being canalized as was first proposed. Completed in 1757, it was to be the forerunner of many other such schemes.

During the latter part of the 18th century and the first half of the 19th, canal engineers, not content with the prosaic form of lock, experimented with lifts and inclined planes – thus fore-shadowing modern Continental practice; the sole working example we still have, the Anderton Lift, has not long since passed its centenary. It says much on the paucity of our modern canal practice that the 29 locks comprising the Devizes Flight on the Kennet & Avon, which from the day of building suffered badly from water shortage, are to be replaced by another 29 locks – what a chance is missed to show what we could do!

Locks, inevitably, reflect the whims of their original engineers, the contractors and modern maintenance methods. To a lesser extent they also reflect the availability or otherwise of cash during the original building of the canal. Rather odd things also happened in the early days which later affected the way the canal was run; among the vicissitudes affecting the famous 56 narrow (70 ft × 7 ft nominal) locks of the Worcester & Birmingham were, and are, the effects caused by the promoters who originally intended to build both a wide waterway and to use lifts; resulting in a rather unfortunate geophysical line being chosen. The original Droitwich Barge Canal had barge or trow locks measuring 71 ft 6 in × 14 ft 6 in with a draught over the cill of 5 ft 6 in, but when the Droitwich Junction Canal was built to connect the barge canal with the Worcester & Birmingham, as this latter only took narrow boats, this in turn was built to 71 ft 6 in × 7 ft with a draught of 4 ft. One of the great 'might have beens' is that had the Worcester & Birmingham canal been built to plan undoubtedly the Junction canal would have been built 'wide' and as a result trows would have worked through to Birmingham.

Conversely, when the Manchester, Bolton & Bury Canal was being built to the narrow gauge, plans were changed halfway through and those locks already completed taken down and rebuilt to 68 ft × 14 ft 2 in, which, although it had the advantage of allowing flats to work through from the Irwell, precluded the use of standard narrow boats – these being 70 ft overall. Furthermore, the tunnel at Bury had a width of 9 ft 6 in at water level, thus negating the advantages of wide locks! When the Huddersfield Narrow canal was built the engineer settled for a lock-gauge of 70 ft × 7 ft but the connecting waterway at the eastern end of the navigation, the Huddersfield Broad canal, was built to take keels measuring 58 ft × 14 ft; thus traffic could only be carried in miniscule boats measuring 58 ft × 7 ft (17.68 × 2.13 m), with an average payload of 15 tons. As, at a time when weekly wages rarely exceeded 40p, transhipment cost 2½p per ton, it will be readily understood how impractical the whole set up was – all the more so when wide boats of 50 ton capacity were in use on their rival, the Rochdale canal.

A final example will suffice to elucidate the quandary of a shipper. The Thames & Severn canal had two gauges of locks and bridges, very large from Stroud to Brimscombe in order that trows could penetrate thus far into the hinterland, and large for the rest of its way to the junction with the Thames Navigation at Inglesham. It was in the end to rue this connection for the Thames was still fitted with, already anachronistic, 'flash' locks and comments from barge-owners, shippers and canal authorities alike were manifold in expressing their disgust at the ensuing delays. When, in 1802, R. Mylne, the engineer, was called in to survey the Thames, he commented repeatedly on the lack of water. At Days Lock, for example, 'In the cut below the lock there were 3 Thames & Severn bargeboats aground for want of water to get into it'.

Below the Thames & Severn was the Kennet & Avon which,   57

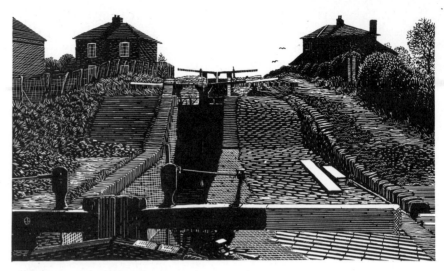

Grindley Brook Staircase (3-rise) Locks. Welsh Branch, Shropshire Union Canal.

although having smaller locks than both the rivers Avon and Kennet, could take 55-ton barges throughout. Joining the Kennet & Avon with the Thames there was another navigation, the Wiltshire & Berkshire canal and later, by a branch (the North Wiltshire) it, in turn, connected with the Thames & Severn. The whole of the Wilts & Berks, plus its branches, was built to the 'midland standard' of 70 ft × 7 ft. Once again through barge working was impossible.

The main reasons put forward by engineers for the continuance of this narrow size in locks being used, were: they used less water, they gave connections throughout the country, what was good enough for Brindley in 1760 . . . More succinctly, they were exactly the same reasons as for the propagation of 4 ft 8½ in gauge railways – cheaper to build, cheaper to run, and anyway they were there. As when Brunel's Great Western broad gauge lines were destroyed, the advantages to the public were overlooked. Even as early as 1795 we find a plan proposed for the Grand Commercial Canal (replacing the Trent & Mersey, opened with narrow locks in 1777) for which the main benefit claimed was: '. . . the forming of a communication for larger boats (40 tons) than the Trent & Mersey is calculated for, except below Burton, and contributing towards the wished for passage of large boats between Liverpool, Manchester, Chester, Hull, London, etc.'

Basically a lock consists of a box with one or two doors at each end, with holes, closed and opened by shutters, to let water in and out. It is axiomatic that being on canals, neither the doors (gates), holes (culverts) or shutters (paddles) are necessarily the same – not only do they differ in all their details from canal to canal but from lock to lock.

58   The gates may be built in wood (greenheart, oak, elm, pitch-

Cerney Wick Lock and Roundhouse. Thames & Severn Canal.

pine, jarrah, even teak) or in steel (framing, or planking, or both); the paddles, usually elm but sometimes oak or steel, can run in wooden or iron slides vertically, or horizontally, or skewed, while the water may enter via a hole in the gates, through a culvert to one side (exhausting into the lock itself or via a weir) or over a weir (which can be open, covered, or be a vertical tube). After those major structural differences the minor details proliferate; the only thing certain is that 'nowt will fit owt else'!

But therein lies one charm of our waterways, and I, like many to whom the canal is a living body, view with foreboding the efforts of those soulless, conditioned beings, who want every detail to be standardized – either for economy (illogical, when new may be bought to replace sound old) or because a few pleasure-boaters accustomed to box-houses and box-cars want it so (insensitive, for is not the old adage 'variety is the spice of life' still true?) or because, the unanswerable argument, it makes a job/money for someone.

Remember railways? When they were different, long ago, the men took pride in 'their' company, whether a giant – the magnificent London & North Western Railway – or an efficient 'little 'un' – the London, Tilbury & Southend. So it was with canals, no one would deny the engineering of the Shropshire Union main line, but, to a workman, say nothing rude about the Oxford! Now 'the Company' has gone for both, instead a computer inspects the track!

70 When this scene was engraved by E. Francis in 1828 it depicted a waterway, the Wilts & Berks Canal, at the height of its fortune, although in truth it was both a quiet enough spot and a poor fortune!

71 *left* Technically a poor photograph, nevertheless it is included partly because of the craft, a wherry, partly because of its age, 1901, and partly because it depicts Honing Lock No. 1, North Walsham & Dilham Canal before it fell into disuse. Although this lock only measured 50 ft × 12 ft 4 in (15.24 × 3.76 m) it looks more impressive than many larger. *L.A. Edwards*

72 *opposite, top* Tyrley Lock No. 1, Shropshire Union Canal in pre-war days. The motor boat (above the top gates) is towing the butty 'overgate' to reduce labour. The girl was, I fear, shy — but it may just have been windy. *Sir John Knill*

73 *opposite, below* On the Lee & Stort Navigation at Sawbridgeworth, the lock-keeper winds the paddles while a (then rare) trip boat waits to enter. Photographed in the 1950s, this scene is already of historical interest, if only in that the length of the girl's skirt looks less fashionable to us now. The gates are remotely activated. *L.A. Edwards*

74 This is the steerer's eye view as a motor boat pulls out of a lock on the Grand Union Canal, snatching its butty behind. In these wide locks the pair are able to work through side by side, by contrast with the (normally) tedious bow-hauling required when traversing single narrow locks. *Sir John Knill*

75 and **76** At the time this photograph was taken (1966) the notice (*right*) pinned to a gate was self-explanatory. Now, as long as it is daylight and water supplies are adequate boats busily bustle to and fro. Marple Locks, Peak Forest Canal.

77 *Below* Stewponey Lock, Staffs & Worcs Canal, *c*. 1920. The boat to the rear of the lock gate has just discharged its load. Quite who is what among the people assembled at the scene is unknown, the gentleman on the left with the bowler hat is clearly a Gentleman, while it is reasonable to suppose the plump lady is off the boat. *Dudley Public Libraries*

78 Succinct indeed is this notice of 1948. At the time when the enamel was still fresh, entry was made through this lock to the (Somerset) Avon, and from thence to the Kennet & Avon Canal. Intrepid voyageurs could then still pass through this route to London — the alternative via the Thames & Severn having been dead for many a year — but now, despite strenuous endeavours by the Kennet & Avon Canal Trust, this way too is impassible.

79 *Below*   This lock on the Rufford Branch (giving access to the River Ribble) near Burscough is both melodramatic and yet typical. The horizontal paddle gearing is just visible on the nearest balance beam.   *L.A. Edwards*

80 *Right*   The roar and tumult of water as a gate paddle is lifted is a sound that cannot be expressed in words. Staffs & Worcs Canal at Bumble Hole.

81 Greenberfield, on the main line, blends neatly into the surrounding countryside late one October evening. The weir (left foreground) flows in a conduit parallel with the lock, to emerge below the tail.

82 Bingley Five Rise Locks are one of the Seven Wonders of the Waterways. The level is raised no less than 60 ft (18 m) in little more than 60 yards (55 m), and gives access to a lock-free pound of no less than 17 miles (27 km). Photographed in the 1950s prior to the recent 'facelift'. One boatman, seeing the unbelievably clean and fresh paint sourly asked me where 'the plastic gnomes were'? *Plus ça change!* *L.A. Edwards*

A canal upon which both amenity and commercial traffics can co-exist is the Leeds & Liverpool, sole remainder of three trans-Pennine waterways. Attractive both in appearance and ease of working, save only in the cities, nevertheless by virtue of the various engineers and contractors employed during the overlong time it took to build (1770–1816) it is somewhat idiosyncratic. The swing-bridges, too, can be a nuisance for the one or two-person crew.

Variety is the spice of life they do say, and around Barrowford Locks one can see many fascinating things 83–87

83 Horizontal paddle-rack and rocking paddle.
84 Vertical screw paddles.

85 One of the stumps. Ostensibly for mooring boats they are usually occupied by uncommonly surly fishermen.

00 'Barrel-organ' paddle gearing.

87 Footbridge; the dog, a Collie, shows the width.

88 *Right* Guillotine locks are nice peaceful things, belying their name. This, no longer operative, version is sited at Kings Norton on the Stratford-on-Avon Canal but rarely do cruisers realize it was once a control lock, designed so that every boat passing from the Stratford paid a lockful of water (that vital canal commodity) to the Worcester & Birmingham Canal. Although the difference in water levels was only 2 in (5 cm), in a year's working the cubic volume can only have been an embarrassment. Work commenced on adjusting the levels in 1959, weirs on the Stratford being lowered subsequently.

The other side of the canal coin would be depressing were it not for the fact that one day, perhaps, each scene will be restored – rejuvenated – and once more bustle with boat-people, whether working or playing. Just how, when, and with whose money is unknown.

89 On what is left of the Huddersfield Narrow Canal locks (technically 'cascaded') mill-girls play at lunchtime. With only a trickle of water passing over, what better than a piggy-back ride – or even a paddle?

90 A number of locks on the Chesterfield Canal are not cascaded but weired. These, at least, will be easily restored when the time comes, for the stonework is as solid as on the opening day in 1778 when 'bands played and crowds came from miles about to gaze at this symbol of man's ingenuity'!

91 The most beautiful location upon the most loved of all waterways. In the meetly named Golden Valley on the line of the Thames & Severn Canal, the left-hand gate is dated 1912; alas the water pours underneath now, a pipe runs across the channel and the bridge is but a conduit.

The Bridgwater & Taunton Canal, running between Bridgwater docks and Taunton, is a relic of a far more grandiose scheme to build a canal across the toe of England, leaving Devon and Cornwall as a separate island. Truncated to a mere route from Bristol to Taunton, this in turn was shortened to the present line. Engineered by John Hollingsworth, between Bridgwater Docks and Firepool, Taunton it has an odd lock size, 54 ft × 13 ft (16.5 m × 4 m) and a shallow draught. Nevertheless it served its purpose well enough until the Second World War when as a part of the anti-invasion scheme the swing bridges had their mechanism dismantled, while pill-boxes were built along the towpath. Rather ineffectual attempts have subsequently been made to re-open it; but as always road users have taken priority and agreement has been reached for a mere 3 ft (0.91 m) headroom to be provided throughout. During the period of procrastination, Bridgwater Docks were closed, so this is now truly a stillwater navigation.

The following three photographs are scenes at Standard Lock:

92 Counterweighted bottom-end paddle-gearing, concrete balance beam.

93 Top end paddle gearing. The counterweight, which greatly eased the drawing of these paddles, is clearly visible, as are bird droppings – they care little that the waterway is still!

94 Bottom end again; the weight is off its gear and uselessly draped across the gate.

95 This ground paddle at Walbutt Lock, Pocklington Canal, has clearly seen better days. Bereft of its safety catch (and oil) only the birds were present on a rough day in 1969. However, matters are in hand and the Pocklington Canal Amenity Society hope that before too long this, surely one of the most serene of canals, may re-enter into full use.

96 Stark against the skyline is this set of paddle gearing and the handrail at Ryders Green Locks, Walsall Canal, one winter's day in 1972. Virtually a standard pattern, this design is to be found throughout the 115 miles (185 km) or so of the waterways that are known as the Birmingham Canal Navigations.

97 Far older are the origins of the Exeter Ship Canal. Generally it is accepted that this is one of the first waterways in Great Britain to be equipped with pound locks in lieu of the old 'flash' patterns but even in 1900 decay was apparent; seemingly the navigation has stood still since its rebuilding under engineer James Green in 1829.

This form of paddle-gearing, nicknamed, reasonably enough, 'Daleks', is to be found on the Grand Union Canal between Braunston and Knowle and was the most up-to-date possible at the time of installation in the 1930s. Although manually operated, providing it is well greased, the mechanism is easy to wind, even if an excessive number of turns seems required.

98 At Knowle is a slightly modified variant; the black object on top of the 'Dalek' serves as a visible guide to the position of the paddle, without this it is necessary to peer through a keyhole. The two upright sets behind control the flow of water to a water-saving side-pound. Using these ensures the water used when lock-working is not entirely lost. They are, at present, disused.

100 *Below* A trifle wintry is this scene at Radford. 'Scotches', giving a good grip to the feet when one is leaning on the steel balance beam, are sharp and clear.

99 The Calder & Hebble still possesses gate paddles of a mediaeval pattern, only one stage removed from those used on flash locks long before Brindley (and Smeaton, the engineer of the Calder & Hebble) came on the scene. Operation is simple, the spike is placed in the slot and levered, one notch at a time, raising the paddle accordingly. Primitive but virtually foolproof.

Lockgates are attached to the lockwalls at the top by a collar plus either 'claws' or an anchor. The bottom of the gate has either a pin dropping into a socket set in the bottom of the lock (or the reverse!) or socket both at heel and in the cill, with an oversize ball-bearing taking the load. Needless to say the fittings at the top vary from lock to lock!

101 Pigeon's Lock, Oxford Canal, 1966. Made from local soft red brick, decay was only superficial in both lock and bridge arch, although a lack of paint made the whole rather sombre howbeit more natural than the 'jazzed-up' locks of today. The uncommon steel balance beam is distinctive.

102 Relatively modern is this system of plate and 'horse-shoe' held by a shaped steel bar and nuts at the far side of the horseshoe.

103 Antique is this arrangement of collar and claw. The collar is held by cotters and spacers ('donkey shoes'), the claws being set in lead, with upwards of 2 lb of lead being utilized to retain each set.

104 The Chester Branch of the Shropshire Union Canal, seen here in 1960, is more often than not quiescent. Many pleasure boat users are daunted by the size of the locks and the weight of the paddle-gearing. Above the lockhouse stands four-square, as it first did in that far off day (September 1779) when the Chester & Nantwich Canal was declared, between 'huzzahs and the firing of cannon', to be open for traffic. *L.A. Edwards*

105 The sea lock at Grangemouth, Forth & Clyde Canal, at a low tide. Not only was it a low tide physically, but a low tide in the affairs of men and women when, under the 'Forth & Clyde canal (Extinguishment of Rights of Navigation) Act', which became law in 1962, this waterway was abandoned. As built the waterway was designed to save almost 400 miles (150 km) for a vessel working around Scotland; furthermore, the provision of swing bridges throughout ensured that any craft could pass through its 35 miles without striking their masts, although it was this provision that, above all, doomed the waterway in this day of the lorry. The cost of elimination is £750,000 for each mile of the canal; and still 'waffling' continues — if a fixed headroom of 14 ft (4.27 m) were accepted there is not a doubt that this desirable route could be re-opened at far less than this cost. *D.G. Russell*

CHAPTER FOUR

BRIDGES

BRIDGES are designed to function. The function is to provide an easy passage through, over or around a depression, however that depression may be formed. Thus they are variously known as tunnels, aqueducts, viaducts or just plain bridges. Originally designed solely as footbridges they were later extended 'to accommodate the traffic of the district'. This gave rise to the later cognomen 'accommodation bridge' – the type most commonly found on canals, connecting perhaps two farmers' fields or carrying some quiet lane. In a sense this term must also include motorway bridges.

As a safety measure most bridges were also given a parapet and this remains in situ on the majority of waterway bridges; not all – for some footbridges, either designed to protect a (now long-lost) right of way, or merely used to cross over the tail of a lock, have never been so provided. In addition vandals finding a loose brick or piece of iron angle soon settle down to dismantle the rest of the parapet – within three hours on a sunny Sunday afternoon a part of the red-brick bridge at Kings Norton Junction (Stratford/Worcester & Birmingham canals) was so treated, requiring no less than 400 bricks for repairs on the Monday.

Other forms of canal bridge are as varied as the colours of the cut. Crossover, or roving, is a word used to describe a bridge which enables the towpath of one line of a canal to cross over the line of another; not necessarily at the junction of two canals but, perhaps, over a factory feeder itself no more than a boat's length. A walk along the main line of the Birmingham Canal Navigations could, until relatively recently, have been likened to walking along a switchback so thick and fast did these feeders come.

Confusingly, a roving bridge can also be one where the towpath swings from one side of the waterway to the other. Known also as 'changeline' bridges, they occurred for a number of reasons – the matter of fact, in that an extant building, mine or similar obstruction, prevented continuance of the path; a matter of cash, the landowner through whose land the cut was dug may have been vehement in his refusal to let 'dirty barges, the pariahs of England' go too near his house and the canal company may not have had sufficient money wherewith to persuade him to amend his beliefs; or it may have been a matter of humanity, inasmuch as the wear on the shoulders and hoofs of the horse, caused by the offset tow, would be equalized.

A particularly magnificent form of bridge is an aqueduct. Disregarding the spectacular – Pontcysyllte, Oldknow, Barton – the mundane invariably represent considerable feats of engineering. Regard the masonry, clay puddle bed and water that make up the channel at Dundas (Kennet & Avon), Brindley's Bank (Trent & Mersey) or Aberdulais (Neath) and imagine the static weight – not transient like that of a railway or negligible as it is on roads – that had been thrust upon the arches and foundations of these works for nearly two hundred years. It is habitual for many modern day commentators to deride the efforts of canal engineers who built these structures, pointing out, rightly, that Telford successfully experimented with iron at both Longdon and Pontcysyllte. However, not many engineers had either ability, knowledge, facilities or finance to risk this kind of enterprise, such was the demand for their services and so great was the shortage, that (unlike today) competent engineers were hard to come by. In any event many were ex-masons who had risen to the rank of 'engineer', often first being a contractor, and being masons they understood this method of work, iron being left to the theoreticians. Josiah Jessop, indeed, originally anticipated using a masonry aqueduct in lieu of Pontcysyllte, having either to bring the canal to a lower level by locks, or to make the channel more circuitous. Chirk Aqueduct, nearby and much lower rated than Pontcysyllte, is a hybrid, part masonry, part iron.

Minor and often overlooked aqueducts are those classified loosely as 'culverts' – they need be no more than a 'saddle-back' land-drain, perhaps only 12 in (30.48 cm) in diameter, or they can give more than adequate headroom for six men walking abreast passing completely under the width of the canal, some 40–80 ft (12–24 m). Whatever their size they nevertheless carry the full weight of the canal, its bed, towpath and banks – and rarely are they inspected or receive any maintenance even when this is practicable, for with older lengthmen retiring or dying the existence of many culverts may be unknown.

Tunnels are among the least conspicuous of all canal features – mainly due to the approach cuttings being dank, dark and dismal. This dissuades many towpath walkers from exploring too far and it is also true that, generally, their portals are anything but exciting. Standedge (Huddersfield Narrow) set against its Yorkshire background of dark stone, looks like a rathole, at least at its Marsden end; the other (Diggle) looking incongruous amid a modern building site. Saddington (Grand Union, Leicester Section), Braunston (Grand Union, Main Line), Wast Hill (Worcester & Birmingham), Harecastle (Trent & Mersey) and many others merely blend into the countryside. In all there were eighteen tunnels on the waterway network that exceeded 1500 yards (1372 m) in length; and a further seventeen over 500 yards (457 m). A new one has recently been built on the Birmingham Canal Navigations Main Line, initial assembly being from reinforced concrete rings; topsoil, rubble, old cars and such junk were then dumped on top to make up the level. In contradistinction to this, at least three tunnels, Brewins (Dudley, No. 2 Line), Fenny Compton (Oxford) and, most recently (1971), Armitage (Trent & Mersey) have been opened out, all due to difficulties in maintenance. The necessity for tunnels proved unfortunate for

Roving or crossover bridge. Shropshire Union Canal.

many canal engineers – those on the Kington, Leominster & Stourport (Sousnant, Pensax, Putnal Field) may well have proved prohibitively expensive for the shareholders; and their canal was doomed never to be completed. The digging through Blisworth Hill was so protracted that the proprietors of the Grand Junction built a tramroad to bypass it, and might well have left it at that had they not been prodded by shippers who objected to the resulting delays, while the completion of the Trent & Mersey was held up for many years by difficulties at Harecastle. It is a salutory reminder of the paucity of our own expertise that repairs to the 'new' Harecastle tunnel (completed 1827 but troubled by subsidence) have continued for many years, culminating in complete closure from 1973 to 1977.

Bore dimensions of tunnels vary enormously, the tub-boat canals of Somerset having the smallest, while those on the Kennet & Avon in Bath, however short they may be, have a superbly spacious appearance. The latter are among the most ornamental as once befitted that beautiful city (beautiful no longer, alas, despoiled as it has been), while Sapperton (Thames & Severn) and Brandwood (Stratford) have attractively worked entrances.

Whatever the form of bridge, one wonders whether it is better to have the superficial 'tarting-up' with a lick of paint as practiced now, or more realistic maintenance of the track – bare iron ribs and worn brick arches are all too common – the first may appear rather attractive, but only the latter is enduring.

Pleasure can always be found when viewing cast iron bridges. They represent a lost art – that of empirical measurement (they look right, therefore they are). Few drawings were ever made; if strength was doubted, everything was doubled.

Aqueduct over River Goyt. Marple, Peak Forest Canal.

'Cock-up' bridge. Welsh Branch, Shropshire Union Canal.

109 A superb example of ironfounders' art, this iron bridge over the Newport Pagnall branch of the Grand Junction Canal is depicted in 1819. Inevitably, an angler pursues his hobby — the water must have been reasonably clear for the horse to be drinking but I fear that when the occupants of the houses behind cried 'Gardez l'eau' in the morning it had to go somewhere!

110 This bridge which once carried the Neath-Abernant Tramroad across the River Cynon at Robertstown, Aberdare, bears the date 1811, when it was cast by the Abernant ironworks. As far as is known it is the oldest iron tramroad bridge still in situ. The cross pieces carried the tramplates. *G. Davies*

111 Wyrley & Essington Canal. Roving bridge at the junction with Lord Hay's Branch Canal. The ribs and rivets show clearly in this early dawn photograph.

Some, if not all, ironworks were proud of their art. Horseley had the advantage of canalside loading facilities, enabling their quotations to be more than competitive. A long-lived firm, certainly dating from the early 19th century, many of their bridges bear the same date; it seems, having made a mould, they were unwilling to alter it.

112 Detail of parapet; unnecessary ornamentation perhaps — but what a delightful fusion of art and engineering.

113 Detail of the skew crossover bridge at Woodside Junction, where the Dudley canal met the Two Lock Line.

114 Staffordshire stone and iron. Caldon Canal, Hazlehurst Junction, in 1969. Then technically closed — despite appearances our boat got this far — the canal is now wholly open to navigation.

115 A swing road bridge. A true 'industrialscape', Sharpness Docks, the terminus of the Gloucester & Berkeley Canal, look quiet enough one December's day in 1965. Originally designed to by-pass a very tricky part of the Severn, this now rather euphimistically named 'Ship Canal', suffered from inaccurate estimating of costs, inflation and generally poor management during its building. Commenced in 1793 — Thomas Telford as engineer to the Exchequer Loan Commission arranged for £160,000 to be made available between 1817 and 1826 — the whole of its modified route from Gloucester to Sharpness was opened in 1827.

116 Butty boat Uranus passing under the roving bridge, yet another ironmaster's product, at Hawkesbury Junction, where the Coventry and Oxford Canals communicate. Taken about 1950, the pumping station, visible behind Uranus, was still in use. *Sir John Knill*

117 A fixed two-part or 'split' bridge on the Stratford Canal. This halcyon scene is at Yarningdale on that part of the waterway not under the aegis of the British Waterways Board. In fact this canal had the nearest escape that any has, or now could have, from being totally abandoned. The story, simply, was that the Great Western Railway, when they owned the route, neglected it and the British Transport Commission were quite happy to let it go. The Warwickshire County Council wanted to lower a road bridge to save money but various canal 'multi-usage' bodies, including the Inland Waterways Association and the Inland Waterways Protection Society, stopped that. Volunteers dug, hacked, chewed, cursed and bullied the canal back into shape. Now (and forever) it is open.

118 Cast iron — but different! This roller, which stopped rolling all too soon, shows the wear and tear of horse-drawn boat-lines. Designed to protect the brickwork of this Birmingham Canal Navigations' bridge, it has clearly been successful.

119 This now old fashioned (for they have become metric) Ordnance Survey bench mark was found on the Worcester & Birmingham Canal in 1967.

120 Cast iron is this bridge plate. Once all bridges bore either name or number — or both — but, alas, these are an all-too-easy to thieve item and the originals are rapidly disappearing. In some areas the British Waterways Board are (to their credit) putting up facsimiles; on the Monmouthshire Canal the Newport Canal Preservation Society have undertaken a similar operation for those bridges in their area.

121 Anderton lift connecting the Trent & Mersey Canal with the River Weaver has a rise of 50 ft. Engineered by Sir E. Leader-Williams (later to be engineer-in-charge of the Manchester Ship Canal) and completed in 1875, it was originally hydraulically operated, using one caisson as a counterbalance to the other, although the caissons are now independently electrically controlled. View from above, looking towards the tanks, with I.C.I. factory opposite.

122 Tern (Longdon) aqueduct on the Shrewsbury Canal, designed as a prototype for Pontcysyllte, succeeded an earlier work which was destroyed by floods in 1795. Disused from 1944, the ironwork is due to be dismantled. *D. Sumner & A. Martin*

Aqueducts vary enormously in their size, if not strength. Apart from the sheer weight of water bearing on the foundations, the force exerted when this is frozen would test any workmanship, while even the passage and hence water compression caused by boats must be allowed for.

123 Engineered by John Rennie, Dundas Aqueduct was virtually de-watered in 1968. The masonry work has charm, although the ornamental balustrade no doubt added to the expense. *J. Russell*

124 Very impressive indeed is Stanley Ferry Aqueduct carrying the Aire & Calder Navigation (Wakefield Section) over the River Calder and opened in 1839. *R. Payne*

125 Edinburgh & Glasgow Union Canal (Scotland), south side of the aqueduct across the river Almond; traditional and elegant. Built entirely from masonry and engineered by Hugh Baird with advice from Thomas Telford, it was first brought into use in 1822. *D.G. Russell*

126 Taken from the bed of the River Neath, this photograph shows both the columns of the railway viaduct and Tennant Canal aqueduct, at Aberdulais. Engineered by William Kirkhouse in 1824, the latter sturdy structure was 'old hat' even then, for the weight of the masonry, puddled clay and water meant excessive foundations and high cost.

Canal engineers, in the main, were allergic to tunnels but the landscape of Britain pays little heed to our whims and perforce they had to blast their way through. In at least one case, the Kington, Leominster & Stourport Canal, cutting tunnels finally brought about the ruin of the scheme; and in another, the Huddersfield Narrow Canal, the very length of the hole (5415 yds/ 4952 m) and the time taken to 'leg' through eliminated traffic.

127 Lillesdon Tunnel, Chard Canal, the western entrance in 1967. Used by tub-boats measuring 26 ft × 6 ft 6 in, propulsion was by manpower. Opened 1842 and closed 1866, the tunnel now serves a thrifty farmer to house his hydro-electric unit.

128 An inside out tunnel, some 100 yards (91 m) or so in length, this is one of two built 1974/5 over the old and new main lines of the Birmingham Canal Navigations. The basic shell is shown here, and then the level was made up with all sorts of rubbish (ostensibly soil) to allow the building of a new by-pass. To the best of my knowledge this is the only time a canal tunnel has been built this way. On the New Main Line Thomas Telford, the engineer, built a waterway with a towpath on both sides; this new erection makes no such provision.

129 The vintage of this illustration will probably never be known as the use of horse-drawn boats through Braunston Tunnel on the quondam Grand Junction Canal did not cease until the 1920s. The load on the nearest boat looks rather irregular. The horses would be led over the hill and, at least from the 1870s, steam tugs acted in lieu through 'the hole', running at a fixed timetable, leaving this, the northern end, at two hourly intervals between 0600 and 2000 hours. *Sir John Knill*

130 The three primary forms of bridge-building are the orthodox brick, cast or wrought iron and stone. Of these three, the mason's art is the most individual, inasmuch as the mason alone could make it seem, as well as be, right.

A particularly fine example of artistry both in workmanship and in design is still to be found at Cosgrove on the Grand Union Canal, Solomon's Bridge having been built there in 1800. *British Waterways Board*

131 *Above* Of equal charm, if more orthodox in shape and less visually congested is Ladysmith (or Ladies) Bridge on the Kennet & Avon. So proud indeed were the workmen that masons' marks are still visible (*see next page*). *J. Russell*

93

132–134 Masons' marks cut in bridge. *J. Russell*

135 Cromford Canal. Simplicity in Derbyshire stone near Butterley.

Some brick bridges show an appreciation of their original purpose inasmuch as the radius would end some 10 ft (3 m) inside the buttress. This apparently square-ish arch leaves sufficient room for both boat and horse. A hoary 'Black Country' tale tells how one day Enoch came across Eli busy cutting away the arch of a bridge. 'What' he asked, 'is yo a-doin? It ay be much easier if yo dug up the towpath.' 'Doe be so saft,' grunted Eli, scarcely pausing. 'It ay the 'orses 'oofs as wo go thro', its 'is yed.'

136 Westport Canal dreams of past days. Somerset stone in tones of honey and rust. Near Westport.

137 Eli's bridge was probably something like this one at Sandiacre on the Erewash Canal, near its junction with the (late lamented) Derby Canal. Pity the poor horse who had to duck under there!

138 In 1838 a tunnel, 75 yds (68 m) long, known as Brewins was included in the new (shortening) line of the Dudley Canal. Named after a gentleman who was both a shareholder and the Superintendent, this was in turn superseded some years later by a bridge also best recognized by the name 'Brewins'! The old tunnel lining may be seen forming the approach walls.

139 Worcester & Birmingham Canal. Astwood. Fixed one piece bridge with slot under at towpath side, built to protect an existing right-of-way and subsequently retained.

140 *Right* Modern architectural engineering is not entirely incapable of producing striking, even aesthetically attractive works. Motorway bridge over the Birmingham Canal Navigations Main Line with Steward aqueduct behind.

CHAPTER FIVE

BOATS

SENSIBLY enough, canal boats are designed to fit the size of the locks utilized within the waterway(s) they are most likely to traverse. This size alone will show how well, or how little, a canal has been modernized during its lifespan – on the bulk of waterways the 'common-or-(lock)-garden' nominal 70 ft × 7 ft narrow boat remains the sole means of transporting goods. That it has been motorized, first by steam, then outboards, later semi- and true diesels is irrelevant; until quite recently the economics of such mechanical power by comparison with equine tonnage were of dubious advantage for capital cost was higher, breakdowns more frequent, and there was a not inconsiderable loss in carrying capacity. The primary advantage of motorboats was that they could work for longer hours without attention but this was to the disadvantage of the boatman for he also had to work longer hours. The unpowered butty (or dumb) boat towed by the motor was crewed free of charge to the management by the boatman's 'mate' (this word used, incidentally, for a steerer of either sex) as only the boatman was paid; meeting the cost of his 'mate' from his own money. At their best, working the Grand Union after the 1930s rebuilding, paired boats were an economic proposition, for they could travel breasted-up (side by side) through the new 14 ft (4.26 m) locks. At the very least the lock-working was halved by this method. In the older single locks on the run to and from the collieries, where pounds permitted, the motor towed the butty 'over-gate' – unfortunately these circumstances were rare. Ideally, had all pounds (the length of water between locks) been equal and more or less straight, no problem would have arisen – but, alas, the ideal was rarely there and the butty would have to be 'bow-hauled' using the mate's labour, too often along a muddy, filthy, ash track, barely pleasant in summer, abominable in winter's rain or slush.

On the broad canals, the Liverpool & Leeds for example, the physical problems were identical; although the payload per boat, 55 ton for a barge in lieu of 25 carried on narrow canals, made more economic sense. Even so the only truly viable waterways today are the Aire & Calder together with its subsidiaries. Viable in the sense that not only is the ton/mile rate more than competitive compared with road transport, but the owners can still quote low rates and be able to invest in modern craft and loading and discharging arrangements. Nevertheless, it is unfortunately all too true that for many years now successive Governments have been wedded to road interests, with canal modernization schemes coming very low down on their list of priorities; indeed such schemes appear to be an anathema to them.

Whereas even two decades ago a living could be made on narrow canals, at whatever cost to men, women and their boats, this is becoming almost impossible unless the steerer uses the

Boatman's 'mate' and traditional butty. Coventry Canal.

money gained merely to supplement his social security.

It is accepted that the use of waterways for amenity purposes is a commendable outlet for you, me and those who can afford it; but who shall take priority at a lock if – as has been projected – a barge carrying 350 tons hoves into sight at, say, Berkhampstead? With the ever increasing congestion on the better known and easier to navigate waterways some pleasure-boaters are beginning to explore further afield, followed inevitably by those who have a vested interest in hire boats, boatyards and trashy souvenirs (all requiring wharfage) until some day the problems that could arise at Berkhampstead will arise at Castleford. Only two decades ago wharf facilities and moorings for pleasure and commercial users were available almost anywhere; already the former are almost impossible to find and the latter grow harder with each passing week. Obviously, for all there may be some discussion, the Government must come down on the side of the pleasure boater (more votes, more money, more lorries) which in turn will lead to barges being delayed, loads declining, boatmen leaving, until the sorry saga of the sixties, a decade that saw the end of narrow boat carrying, will be repeated on the broad canals. The alternative is to make some agreement now and to stick to it – possibly, if the financial position of the country improves, provision for separate locks of a suitable size might be feasible; when all is said and done a 20 ft × 7 ft boat using a 457 ft × 22 ft lock cannot reasonably be said to be economical use of water!

The evolution of pleasure craft is a fascinating subject on its own. Although on the Broads, Thames and Humber there have always been specialized boats catering for the holiday-maker, nevertheless the nearest to these on canals were the Packet 99

Boats which offered a service rivalling the stage coach, being fitted out pretty thoroughly:

> For here a cabin at each end is found,
> That doth with all conveniences abound.
> One in the head, for ladies nine or ten,
> Another in the stern, for gentlemen.
> With fires and tables, seats to sit at ease,
> They may regale themselves with what they please,
> For all utensils here are at command,
> To eat and drink what'er they have at hand.

When these were swept away by railways which, although dearer, were faster, the hire of standard cargo-carrying boats suitably refurbished was quite a well-known feature of Victorian life. Much favoured for Sunday School outings, anything from 100 to 300 children would be wedged in each, and, horsedrawn, the convoy would make its way out of the grime of Birmingham, Stoke or Manchester into the green fields.

The local canal company's superb inspection boats could also be hired when they were not wanted by the directors. Magnificently fitted (plenty of glass, mahogany, leather) with a couple of handpicked steerers and suitable ales for quaffing, no more pleasant way could be found to see our countryside. A few wealthy Victorian personages had their own boats but at that level of wealth, canals were rather 'non-U'. After the First World War the appreciation of boats appears to have spread from the rivers and the Fens until the hire firms we know today began to make an appearance. One at least, Inland Hire Cruisers at Christleton, already had 13 boats by 1939.

After the Second World War, encouraged by the books penned

Derelict boats. Coombeswood Basin, Dudley Canal No. 2 Line.

by the late L.T.C. Rolt, *Narrow Boat* among them, and that of Eric de Mare, those 'in the know' began to hire in ever increasing numbers, while sales, either of converted narrow boats or, more likely, ex-lifeboats, began to appear regularly in the 'small-ads' columns of water-aligned magazines. Not that even two decades ago hire-boats were very special – the cabin probably leaked, the engine could well break down or get clogged with weeds, re-frigerators, showers, flush toilets were rare indeed – more likely an evaporative cooler for butter, a manual pump and a bucket-lavatory – although a spade was normally supplied for use with this latter. On the other hand no one was in a hurry, the owner probably had a chat, would lay on your food for you, come out willingly when the whatsit broke off the oojamaflick, and it only cost £15 a week for a two berth – less out of season!

Rather more superior boats came in the middle 1960s; albeit they probably still leaked; converted from narrow boats, and often with wooden hulls, they represented possibly the best compromise, requiring a modicum of skill, or at least attention, but were rewarding in their natural feeling and cabin warmth. Today, all is provided, steel hulls, fluorescent lighting, showers, fridges, carpets, luxurious bunks – sorry, beds! – full size cookers, even playpens for the children, but somewhere for all too many hirers, the art of standing still and watching or just ambling along is subordinated to the necessity of doing their set miles per day, and in the evening (having hired the thing) on goes the 'telly'. Not all are like this, those that are not will usually be found on some quiet canal on a spring or autumn evening lean-ing against the cabin placidly eyeing, and being eyed by, an old Friesian cow, probably both meditatively chewing the cud, maybe dreaming of days of yore.

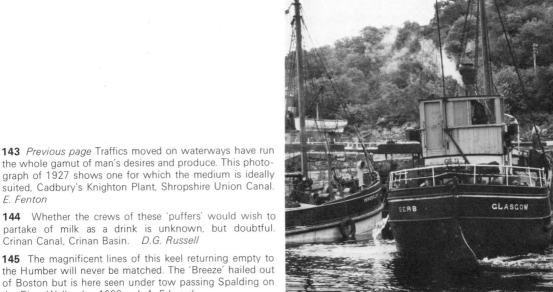

143 *Previous page* Traffics moved on waterways have run the whole gamut of man's desires and produce. This photograph of 1927 shows one for which the medium is ideally suited, Cadbury's Knighton Plant, Shropshire Union Canal. *E. Fenton*

144 Whether the crews of these 'puffers' would wish to partake of milk as a drink is unknown, but doubtful. Crinan Canal, Crinan Basin. *D.G. Russell*

145 The magnificent lines of this keel returning empty to the Humber will never be matched. The 'Breeze' hailed out of Boston but is here seen under tow passing Spalding on the River Welland *c.* 1900. *L.A. Edwards*

The design of the classical narrow canal boat has probably changed little over 200 years. Originally wooden, later composite (iron sides and wooden bottom) or all iron, their final flowering were the steel craft built for the Grand Union Carrying Company in the 1930s by Yarwoods at Northwich and Harland & Woolf at Woolwich. Although the design was in effect fossilized, even today nothing better has been built for the trades they were intended to carry.

146 Two ex-railway butty boats in British Waterways (British Transport Commission) colours. March 1963 at Wolverhampton. 'Kidsgrove' and 'Crewe' are attractively decorated on the doors; and the brasses on chimneys and inside the cabin of 'Kidsgrove' have been well polished. *M. Webb*

147 This narrow boat, seen at Shepperton on the Thames in 1950, has brought topsoil from Newbury (River Kennet Navigation). An unorthodox use for a dustbin! *Sir John Knill*

148 A Samuel Barlow Coal Company boat on the Grand Union in the late 1930s. Fully lettered, the colours of the paintwork made a magnificent pageant. The protection given to the brass chimney bands shows the boatwoman's care of her craft. *The Geographical Magazine, London*

149 Grand Union Canal Carrying Company, No. 208, in wartime livery with economy lettering, manned by one of the volunteer boat-girls. 'Alphons' is a 'small woolwich', the name being a combination of the height of the sides (small, middle and large, according to tonnage capacity) and place of building (Northwich or Woolwich). *The Geographical Magazine, London*

150 Another 'Barlow' boat, but the 'Hyperion' and her butty belonged to S.E. Barlow of Tamworth. Photographed in the late 1940s, the boats are in 'natural' rather than 'tarted-up' condition. *Sir John Knill*

151 The operations involved in boatbuilding were rarely photographed. This yard, Bushell Brothers of Tring, on the Grand Junction Canal, was obviously successful in 1902 with both commercial and pleasure craft on the stocks. The varieties of coal bear witness to their other activity, although doubtless some was destined for the Flour Mills in the background. *Museum of English Rural Life, Reading and Miss C. Bushell*

152 Mainly, one presumes, because the boats damage the road surface. At the rate canals are retrogressing it is as well this notice has been repainted! Photographed in Reading, 1950. *P. Myall*

153 When in 1949 the N.B. 'Columba' carried 20 tons of flour from Ellesmere Port to Stourport, the Staffs & Worcs Canal was in such a poor condition at the bottom end that a horse had to be enlisted to aid tractive effort. The boatman (*left*) is seen with two canal men — and the vital member of the crew! *Sir John Knill*

154 'Linda' (ex 'Victoria') is here seen at Gas Street, Birmingham, in the colours of a postwar carrying company. Photographed in 1967, this is a truly historic scene inasmuch as the basin suffered from 'developer's itch' in 1975, the bulk of the buildings being demolished, an action which received the suggestion in 'Waterways World', April 1976, that it should be entered in the Guinness Book of Records 'as the biggest act of vandalism yet perpetrated'.

155 Anderton, Trent & Mersey Canal, 1966. Motor boat 'Apple' and butty boat 'Bangor', the latter one of the last two butty boats built for Fellows, Morton & Clayton's North-western fleet.

156 The 'Susan' of Alfred Matty & Sons, Coseley, was a typical Birmingham Canal Navigation tug, and like most craft utilized on that system, has an indefinable battered — but busy — look. Netherton Tunnel Branch, September 1952. *W.K.V. Gale*

157 Twelve Lancashire boilers require feeding at the Bilston Iron & Steel Works, c. 1920. While the five on the right were normally fired by blast furnace gas (a thrifty move indeed) the other seven would be capable of burning only small coal. Several boatloads of this coal (slack in canal parlance) would be needed every week, being unloaded by hand to a space in front of the boilers and then thrown into the fire-holes as required, again by hand. This work went on day and night, 365 days a year, both boilermen and boatmen seemingly being bereft of shelter. *W.K.V. Gale*

158 By contrast, an all modern scene in 1975. The canal has been recently rebuilt, the motorway thunders nearby, while pylons stride overhead. Tame Valley Canal, near Rushall, push-tow tug 'Turnpike Sailor'.

159 This tug, 'Parrot' of the British Waterways Board's fleet, differs from both 'Susan' and 'Turnpike Sailor' inasmuch as no living accommodation is available on board. Worcester & Birmingham Canal, Selly Oak.

160 This steam tug of the early 1900s also lacked living accommodation. In 1860 William H. Bartholomew, engineer of the Aire & Calder Navigation, first developed the concept of 'Pans' or 'Tom Puddings' which are tweaked bodily from the canal and upended into enormous floating hoppers at Goole. Tug 14 is pushing the 'jebus' or dummy bow, normally attached to the front of loaded boats where it acts as a cutwater. The whole train is running back to the collieries for loading. *P.L. Smith*

161 Cawood Hargreaves Ltd tug towing three compartment boats in the River Calder after loading at Calder Grove. The restricted depth of the Calder & Hebble Navigation reduces the cargo to a mere three hundred tons in lieu of the normal 'pan-train' load of 500. Near Wakefield, 1970, Aire & Calder Navigation. *P.L. Smith*

162 In this historical photograph, a 'Rotherham' class push-tug passes Ferrybridge Power Station propelling two laden 140-ton lighters to Hull where they will be transshipped aboard their mother-ship 'Bacat 1'. Aire & Calder Navigation. 'Bacat 1' ceased operation in December 1975 partially on economic grounds, although without question difficulties with the dock labour force aided and abetted this decision. The shame of it is that this was one attempt to modernize water transport; for using this container system boats could work through the European canal network. Now only the traditional craft (seen unloading at the power station) will continue to be used – and more lorries will crash through our streets. *British Waterways Board*

163 The 'Susan Peake' of Hargreaves fleet is leaving Bulholme Lock, near Castleford on the Aire & Calder Navigation, laden with 190 tons of coal, October 1975. The railway line ahead is (regrettably) a victim of road competition.

164 Leeds Co-operative Society coal barges lying in Leeds New Dock, July 1971. With the exception of (*right, rear*) one motor barge, all these are unpowered; the Northern version of the traditional narrow canal butty boat. The Co-op ceased handling house coal by water in April 1975. *P.L. Smith*

165 Motor barge 'Lys' at Town Lock, Doncaster, Sheffield & South Yorkshire Navigation, in 1970, en route to Rotherham with 100 tons of grain. The Carley float on the roof is a compulsory fitment for vessels navigating the tidal River Humber. *P.L. Smith*

166 'Shippes and boates go down to the sea'. In the foreground is an Aire & Calder Navigation 500-ton petroleum tank-barge 'Battle Stone'. Behind her another empty tanker passes Hull Docks, while to the right the now-discontinued Bacat service mother-ship 'Bacat 1' makes her way towards Rotterdam. Corys ceased operating tankers on 28th March, 1976, although some of the craft remain in service. *British Waterways Board*

167 In a democratic, classless, society anyone can take to the water — even as long ago as 1920 this happy looking crew took their 'yacht' along our canals. *E. Fenton*

168 A Sunday-school outing on Harry King's boats, photographed around 1910. A rough estimate shows there to have been about 150–200 children and adults; as well, perhaps, that the Board of Trade were not advised. Even today it is not uncommon to find elderly ladies and gentlemen who remember the anticipatory excitement, the sheer fun and the eventual peace of such an outing. *P. Garrett*

169 Although dated June 17th, 1904, save only for variations in the ladies' dresses, this photograph of the P.S. 'Gondolier' at Inverness on the Caledonian Canal could have been taken at any time between 1866 and 1939. Described as a 'handsome and well-finished vessel', she operated throughout this period between Inverness and Banavie, not leaving until 1930 when she went to Greenock for a 'new' boiler — new only to her, inasmuch as it was made in 1902! Refitted with new saloons in 1935, this paddler might have lasted far longer but was requisitioned by the Admiralty, stripped and sunk as a blockship at Scapa Flow.

As has already been seen the use of canals for amenity purposes is nothing new; but it is only since 1967, when such headlines as 'Go Cruising Boost' and 'Holiday Charter for the Canals' appeared in the press that this has become a hobby, however expensive, for men and women of all walks of life. When saturation point will be reached throughout the network is anyone's guess, although the signs are it cannot be far off.

170 Typical of an almost defunct style is this parana pine and mahogany plywood hulled craft, built at Hanbury in the 1960s. Although a number were used in hire fleets, private ownership is more satisfactory if their life is to be extended. Alvechurch, Worcester & Birmingham Canal. *British Waterways Board*

171 This gay waterbus, almost continental in appearance, plies between Bath and Bathampton and in true packet-boat style, by request, picks up and deposits its passengers at any point on the route. *Sir John Knill & Sons*

172 Purpose built is the 'Kenavon Venture', seen here at Bath in 1966. The Kennet & Avon is extremely shallow — as little as 8 in of water has been known and until the entry of the packet-boats 'Mallard' and 'Kingfisher' into service, weed-growth could overwhelm any propellor. Although odd to look at the 'Kenavon Venture' gave, and still gives elsewhere, good service.

173 Unusual was this water carnival at Govilon on the Brecon & Abergavenny Canal in 1967. The Viking craft was from the look of it, transporting Pictish slaves towards the settlement at Gilwern. The rune-cutter was, presumably, under the influence of grog . . . or a slave! *G. Davies*

174 'The Ancient Beldame who inhabited footgear' is followed by the 'Hillbillies'; pleasant occupations for a one-time ice-breaker and a 'railway' boat. *G. Davies*

175 One of Robinson's Hire Cruisers of Dewsbury (Savile Town), the 'Finisterre', a 6 berth, built in 1973, pulls away from the bottom of Bingley 3-rise locks, Leeds & Liverpool Canal. In the foreground is a British Waterways Board maintenance boat. Clearly the hirers have had boats before — lock-wheeling (cycling between locks to make them ready) is de rigeur! *Robinson's Hire Cruisers/R. Payne*

176 Farndon Harbour Moorings, near Newark. A sensible marina site, flood gravel workings were utilized, the whole being connected to the River Trent by a short canal. As the site had already matured, tree-covered islands were retained, thus avoiding the potentially harmful visual impact of a phalanx of craft glittering in the sun. How long such an idyllic set-up can remain is the problem; after all, people want to get out of modern 'people boxes' with their miniscule gardens and out into the open. Where better than on the canal — unless that, too, becomes a seething maelstrom of boats. *British Waterways Board*

CHAPTER SIX

MAINTENANCE

THE CHANGING fortunes of our waterways are nowhere more obvious than in their maintenance. In the beginning, for some waterways, all their manpower was devoted to rectifying the sins of contractors – repairing locks, stopping leaks, planting hedges and generally fettling up the track. In order to get an income it was all too often necessary to allow the entry of water into the bed long before it had properly settled and many were the breaches, bankslips and distortions which occurred. However, assuming the waterways concerned were profitable this meant extreme wear and tear on locks, accentuated by the twin necessities of speed in movement and the lack of expertise of many boatmen – although a few may have been good, such was the expansion in the number of boats that unskilled and uncaring men must have been employed – this latter a problem which has bedevilled maintenance men throughout! Unfortunately for good maintenance, traffic worked a twenty-four hour day and stoppages for gate replacement had to be fitted in as and when possible, demurrage having to be paid for any boat which was delayed above three days. For this reason it became the practice to close the canal completely for a given period, perhaps Easter or Whitsun. Each canal would have a gang in excess of a hundred, every man having and knowing his task, the work being programmed and continuing long into the night, lit by naphtha lamps.

In 1901 the Worcester & Birmingham canal still employed no less than 86 maintenance men, by the 1940s this had dropped to an establishment of 45 which, however, still allowed for five carpenters plus an apprentice, five bricklayers together with a trainee and their three labourers and six lock-keepers in addition to 'navvies' etc. Now, with the Northern Stratford added to the length, only 29 are employed, two bricklayers plus 2 labourers, two carpenters and their mates, four men employed on dredging and two lock-keepers. Discounting the two foremen and the Section Inspector this leaves 12 men for routine maintenance.

Even in relatively difficult wartime conditions (1917) we find on this canal ten men were involved in a stoppage, the gang used then comprising two carpenters plus apprentice, a lengthman, three lock-keepers, two tugmen and one labourer. Major repairs to locks were carried out during the Whitsuntide, but it was normal practice to have a stoppage for a few hours each Monday, thus enabling routine 'fitting round' to be done. It will be noted that these stoppages took place during the spring and summer seasons, mainly to gain the benefit of long daylight hours and better weather conditions. More recently (certainly up to 1952), on the locks at Aston (Birmingham) stoppages were invariably held during the Bank Holidays, there being negligible traffic about as factories were closed.

The effect of any sudden stoppage could be unfortunate, as

Shafting a boat towards Gorsty Hill Tunnel, Dudley Canal No. 2 Line.

when on August 27th, 1949, an emergency works had to take place at Hawkesbury Lock (Oxford canal) '. . . when the top gate had been lifted out of its cup' the lock being drained before it could be adjusted. The duration was from 0930–1600 hours and no less than 25 loaded boats plus ten empties were held up. Notwithstanding a Docks & Inland Waterways Executive note that 'the delay would not make any difference to their loading', this would be a day's pay lost to the boatmen, paid as they were on a trip basis.

In March of the same year an involuntary stoppage took place at Lock 39 on the Grand Union, Leicester Section. The intention was to carry this out during the 'Easter Stoppage' but damage was so severe the work was brought forward. Work done included 'removing estimated 30 tons of bricks and concrete slabs . . . top gates relined, fitted and part planked . . . top paddles dressed [fitted] and the butt [of a tree] removed from culvert'. Twelve men were employed for the week at a total labour cost of £45.05. Winter months were then used in the workshops to build gates – from a log to the finished product – prepare paddles, wooden paddle-starts, dredging and common barrows.

Lock repair is one of the three essentials if a canal is to keep going, another is leak stopping. Basically this process involved dewatering the relevant stretch of waterway, removing the affected bottom for a reasonable depth – if a fissure has occurred   119 this must be plugged with hardcore, concrete or clay, as circumstances require – and then over this a layer at least 24 in (60.96 cm) deep of good puddling clay (usually blue/grey) is worked. At best this is laid in 2–3 in layers and fettled either by cattle hoofs, boots, or by mechanical means until a homogeneous mass is formed, successive layers being added as possible. Normal modern practice appears to lie in pushing clay into the hole by the dumper load, then driving a bull-dozer over it, finishing off with an excavator bucket. If the damage is excessive pre-formed concrete, fibre-glass or butyl-rubber sections may be used; all then being coated with clay.

To close a canal for such works three methods may be employed. Existing safety gates, for all the world like a single pair of lock gates, may be pulled to, although this is easier said than done for the silt of aeons may have obstructed the bed of the canal between their open position and the cill, to which theoretically they abut. This presumes – never a sensible thing to do on waterways – that the cill is either intact or even still in situ. More sensible practice is to use a set of stop-planks which can be dropped down iron grooves set in the brickwork of a bridge or stop-place, although this is fraught with difficulties, for these grooves too may be choked with muck, be broken or so cankered as to jag into the planks. As if this was not enough at 3 a.m. on a wet winter's morning (breaches always occur at 3 a.m.!) on some waterways it is found that the grooves are battered inward from top to bottom and although planks are carefully numbered and allocated they may well no longer be their original length. 'Sometimes they grow, an' sometimes t'Cut bloody shrinks', dourly commented one carpenter after a couple of hours struggling with a particularly recalcitrant set.

On the other hand, due to settlement of the surrounding land or alien pressures, a motorway or similar, the top of the groove may 'come in' or be pinched, requiring all the planks to be cut down (a risky procedure as they can pop out as pressure comes onto them). If the groove is in brickwork it will have to be hacked away, if it is iron, a sledgehammer may be utilized. If all else fails the canal may be stanked off by having heavy duty interlocking piles driven in, although as this is somewhat noisy the natives may take umbrage. When a woman living near one such site came out and started chittering, the British Waterways Board's foreman turned ponderously and told her, rightly, 'Mi lass, thee's got two chances, a yed-ache or a swim. Which doesta want?' Within minutes every light in the street was lit and squawks resounded to the heavens, much to the amusement of the men working, the leak having already been reduced to a trickle.

Normal length piling, using lightweight interlocking sheet steel piles, is carried out with either a boat-mounted jig or a compressed air hammer. If a two-man gang is used the jig is preferable as the piles should keep in line, a third man being added when the job is done freehand. The piles are then faced with a pressed steel strip and retained by means of rods and piles driven into the bank. Prior to this method concrete piles of various shapes and sizes were extensively utilized; ideally these should have a 'toe' or sharp point, and be moulded to allow one to interlock with the other. During the moulding it is desirable that reinforcing rods be incorporated and the sides left straight and true. Alas for the frailty of man, when bonus schemes were in use the rods were often omitted, the interlocking grooves were abandoned and too often they were moved from the mould while 'green' and more resembled a banana than a pile. During boat loading casualties occurred, and after a few breakages due to the driving, if 60 per cent were finally used this was a good figure. These were normally driven by means of a 'dolly' or hammer head raised either manually or mechanically and then smartly lowered. When all were driven, lengths of redundant railway track were bolted on the face, and back ties led to 18 in cubes (45.72 cm^3) of concrete.

To back-fill piling sites, dredging waste is often used, and in this field modernization can have both advantages and disadvantages. For example, when a steam dredger was in use during 1957 a typical week shows there to have been five men employed, receiving a total wage of £72.5.3¾ (£72.2656) between them, hired plant plus other expenses totalled £59.36 and no less

The charm of a canal.

than 423 tons were shifted. Nowadays, when four men are employed using diesel powered equipment, it is rare for a figure in excess of 200 tons to be recorded. Nevertheless, it is still an unpleasant job, recognizable corpses being one of the less salubrious items to be found.

Aside from these major jobs, many are the activities of maintenance men: hedge-brushing and laying, fitting paddles, greasing gearing (although too often now the collar holding the gate is omitted from this process), towpath leak-stopping, ditching, cutting or spraying grass on the banks, weed-cutting, scavenging floating rubbish, painting bits and pieces, putting up notices, titivating BWB yards, dry-docking, loading and discharging maintenance boats, usually with a vintage manual crane, pointing and repairing brickwork, erecting fences, removing fallen trees resulting from storm damage, making good towpaths, water-running to and from reservoirs, repairs to tied cottages and many other occupations in all of which he is expected to be an expert. Finally, we must not overlook that most solemn duty of any self-respecting canal man, whatever his rank – lighting fires and keeping them (the fires, that is) warm!

179 Not, one might think, in the main line of maintenance. It is, in fact, purely coincidental that this man, a carpenter by trade, happens to be an angling enthusiast. When a stoppage, and hence dewatering of the canal, takes place, although not obligatory it is customary for the men on the site to rescue the fish. Regrettably this is frowned on in some areas as a waste of time, but pity the poor strangled fish.

180 And this lot was dragged out of the canal bed with rakes; an unpleasant, stinking, muddy job that has to be done — but rarely do BWB men volunteer! Tardebigge Yard, Worcester & Birmingham Canal, 1972. Section Office to left, the 'Swiss Gnome' cottages house one-time and serving employees.

181 This sort of rubbish, common enough, is far from a joke, although the dog is enjoying a rat-catching session. Sneyd Reservoir, 1972.

182 Far neater is this BWB Maintenance Yard at Marsden near Standedge Tunnel, Huddersfield Narrow Canal. The supply of water for industrial purposes provides almost the sole income for this waterway and maintenance of the track is as vital, howbeit different in certain aspects, as that of a heavily used canal.

183 The British Waterways Board's workshops at Bulbourne, Grand Union Canal, were far from pleasant, murky, dripping and reeking of car exhaust fumes in February 1973. The craft lying under the hoist is a British Waterways Board tug, one of a number built from shortened carrying boats. Behind this is the former inspection boat 'Kingfisher' used by the management in their perambulations around the system.

Silting of the channel has a number of effects. The obvious one is that as the water gets shallower so payloads on boats are reduced. As, in general, the canal has to act as its own reservoir the cubic capacity is reduced leading to restrictions on lock usage. Finally tyres, tree-trunks, polythene and almost any portable items will in due course foul gates and paddles or choke weirs leading to further loss of water and/or flooding.

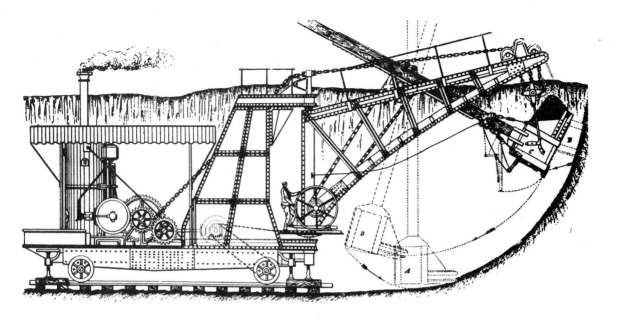

184 This is the sensible way of getting the mud out of the channel. Although an expensive operation and not always practical, due to poor access, it is easier than shovelling and, more importantly, releases workmen for duties elsewhere. Worcester & Birmingham Canal, Lock 42, March 1973.

185 Hy-Mac diggers used on stoppages are really only a derivation from this 'Steam Navvy' used on the Manchester Ship Canal during its building. The Board of Trade might, today, take a dim view of all those exposed whizzigigs whirling about; but in its time it reduced the cost of excavation from the navvies' rate of 5.83p per cubic yard (0.7646 m³) to 1.04p.

186 Unfortunately, we have, in terms of tonnage lifted, retrogressed somewhat from those halcyon days as this machine, seen on the Welsh Canal in 1965, rarely moves more than 300 tons a week at a somewhat higher cost!
L.A. Edwards

187 In the circumstances this notice is, to put it mildly, unfortunate. I will spare blushes by not naming the waterway.

188 At Burghfield on the Kennet & Avon Canal a new lock was built upstream of the old in 1968. Piling is driven showing the outlines of the chamber.   *M. Denney*

189 Even today some work is carried out in situ, thus here a new cill is sawn to size. Power tools are rarely used on site; bulk and economics are against them. Worcester & Birmingham Canal, April 1973, Lock 30. The sawyer to the left, Edgar Shrieves, retired in December 1973 after 50 years service with the Board and its predecessors. We shall not see the like of such men again.

190 In long bygone days locks were built from brick, while the banks were dug out by hand and lined with stakes, stone copings or plain clay. Lockgates were manufactured at either the yard or the local wheelwright's shop. On the Manchester Ship Canal the weight of each gate (230 tons of timber plus 20 tons of iron) required that they were assembled in situ, as here at Eastham.

The work of lock fettling goes on throughout the year, although the more arduous tasks – fitting new gates, renewing brickwork, changing paddles or cutting new weirs – are normally concentrated in the winter months thus causing as little inconvenience as possible to the cruising fraternity.

A rather different atmosphere is generated by volunteer working parties, available person-power is, usually, far greater, but equipment has often to be 'make do and mend'. While safety helmets are de rigeur in the locks on BWB stoppages, their provision is variable in these circumstances.

191 The frost lay thick on everything at 0745 one morning in April 1975. Worcester & Birmingham Canal, Lock 31.

192 Gently — gently. A new 'top-end' gate is tried for size, as on the same day the sun rises, giving an illusion of warmth. Worcester & Birmingham Canal, Lock 30.

193 A month earlier, on a sleety day, the carpenter gives his instructions to the men below, as a 'bottom-end' gate is joggled into place. Worcester & Birmingham Canal, Lock 42.

194 An attempt to hoist lockgates from the Summit Lock on the Rochdale Canal on Saturday, 23rd October, 1975, by members of the Rochdale Canal Society. A superbly engineered waterway (John Rennie, engineer), it was opened in 1804 at a cost of £600,000. This portion fell into disuse about 1952 (the last boat passing just after the war) although the length in Manchester survived longer.

R. Payne

The Pocklington Canal, situated in Yorkshire, is one of the less orthodox to undergo restoration work. Never physically connected with the main circuit of waterways it was, at best, strictly pastoral both in appearance and trade – although it once achieved a weekly packet boat, via the tidal Derwent Navigation to Hull. Sold to the North Eastern Railway in 1847, by the 1950s it seemed its course was run, a proposal being put forward to infill the channel with deactivated sludge but now, under the aegis of the Pocklington Canal Amenity Society (founded 1969), a better future seems assured.

195 York University Conservation Corps volunteers assist Pocklington Canal Amenity Society members to clear the towpath above Cottingwith Lock one spring day in 1975. Pocklington Beck, the primary water feeder to the canal is on the left, and behind this is Wheldrake Ings 'wetland' nature reserve run by the Yorkshire Naturalists' Trust. *Yorkshire Gazette & Herald Series, York.*

196 The dressed stonework of the wall of the Rochdale Canal is to be admired although, after BWB standards, the methods of hoisting lock-gates looks a trifle hazardous, if not inadequate. There appears to be a modicum too much of mud for comfort. *R. Payne*

197 March 1975. Impossibly clean (cf illustrations of British Waterways workmen!), this party of volunteers are repointing and 'fettling-up' brickwork on Hagg Bridge. One of this party is handicapped but thankfully it is not now uncommon to find disabled people working on waterways. This bridge, which carries the B1228 road, is one of the four on the canal 'listed' by the Department of the Environment as being 'historical monuments worthy of retention'.
Yorkshire Gazette & Herald Series, York

198 Photographed 1963 near Inverness. Not, alas, Nessie the monster, but a common or garden diver. A diver on a canal? Well, it is the Caledonian! A fouled paddle was the cause of his immersion. *G. Davies*

199 Shropshire Union Canal, Welsh Branch, breach at Hampton Bank, 1st February 1975. Some idea of the extent of even a simple washout can be gauged from the 2 metre tape being held by the BWB man. The flooding was somewhat extensive. *British Waterways Board*

When originally built the availability or otherwise of cash was reflected in the finish of a canal bank. The straightening works on the Oxford Canal, superintended by Thomas Telford from 1829 to 1834, are lined with stone, the modernization plans of the Grand Union Canal in the 1930s led to many stretches being piled or protected by concrete walling. As motor boating increases, this 'campshedding' takes on greater importance, for erosion not only silts up the track but can lead to breaches of the banks.

200 Leeds & Liverpool Canal at Keighley in 1952. A maintenance boat — possibly even designed for piling — is at least bridging the gap. *J.K. Ebblewhite*

201 Back filling can be in clay, clinker or shale (red ash); this latter being recovered from old mine workings. Loading a boat at Lifford Lane Swing Bridge, Stratford Canal, 1974. This was the nearest point accessible by road — at the site it will be shovelled out.

202 Pontcysyllte on the Welsh Section of the Shropshire Union Canal must be the best known of all canal structures. Designed by Thomas Telford it is '126 ft 8 in in height, from the surface of the flat rock on the south side of the river Dee to the top of the iron side-plates of the water-way . . .', and was completed in 1803 at the cost of £47,069. For the first 160 years of its existence it led a humdrum life, although much admired. The railings were renewed in the late 1960s to the original pattern; and the mortar of the stonework (originally lime and ox-blood) was repointed while the ironwork was given a lick of paint from time to time. Drainage is by way of a bung in the middle and whenever the channel is dewatered the opportunity is taken to clean out the rubbish. On 'Black Friday' the 13th June, 1975, it was found that within the downstream end span two of the inner iron ribs had fractured and cracked and the outer ribs were noticeably buckled. The trough was immediately dewatered and the canal closed for remedial work to be undertaken, not being reopened until February 1976.
British Waterways Board

203 Icebreaking, although no longer practised today, was a normal feature of maintenance work. A very poor quality photograph, this dates back to the first world war and depicts icebreaking somewhere (possibly the Tame Valley Canal) on the Birmingham Canal Navigations. The horses provide the tractive effort, while the men roll the boat from side to side. *Birmingham Central Libraries*

204 When this photo was taken at Knowle in 1935, during the new works of that period — including widening the locks and fitting the 'Daleks' — these men looked happy to be there, for this modernization was undertaken as part of a Government scheme to relieve unemployment. They had the knowledge, too, that the shilling in their pocket would be a shilling tomorrow, the day after and the day after that, not 5p today, 4p and 3p! *P. White*

205 It is understood that the Chancellor of the Exchequer has this photograph on his wall. Taken early in 1963, it is a good illustration of just how hard the great freeze was. This is Bittell Cutting, Worcester & Birmingham Canal, and both the men and the van are standing on the waterway, the opportunity having been taken to trim the overhanging growth. Left is T.W. Mills, the Ganger who retired in December 1974 after 47 years service. Handing him the paybook for signature is George Colledge the then Section Inspector who himself retired in March 1974 after 46 years work, having joined the Sharpness Dock Company in 1928. *D. James & S. Turner*

206 When maintenance boats need moving what better than the old ways? For various reasons only about half the British Waterways Board's maintenance boats are, or can be, fitted with motors. It is impractical to keep a horse available due to the cost of hay, however it is widely rumoured, and accepted, that any canalman will perform well on cutwater, fresh air and a fag. Worcester & Birmingham Canal approaching Astwood Lock, December 1972.

207 At the end of the day a couple walk homewards along the towpath. Although spring was just burgeoning and there was a nip in the air, this lady and gentleman, like many thousands before and after them, were enjoying some of the pleasures a canal can offer: peace, birdsong and a gentle charm.

INDEX TO PHOTOGRAPHS

142